SpringerBriefs on Pioneers in Science and Practice

Volume 52

Series editor

Hans Günter Brauch, Mosbach, Germany

More information about this series at http://www.springer.com/series/10970
http://www.afes-press-books.de/html/SpringerBriefs_PSP.htm
http://afes-press-books.de/html/SpringerBriefs_PSP_SirPeterHall.htm

Richard D. Knowles · Céline Rozenblat
Editors

Sir Peter Hall: Pioneer in Regional Planning, Transport and Urban Geography

IGU
UGI

Springer

Editors
Richard D. Knowles
School of Environment and Life Sciences, Peel Building
University of Salford
Salford
UK

Céline Rozenblat
Faculty of Geoscience
University of Lausanne
Lausanne
Switzerland

The cover photo was obtained by Chia-Lin Chen from *Planning Magazine* that granted the copyright permission to use it in this Springer book to the editors. A book website with additional information on Sir Peter Hall, including videos and major book covers is at: http://afes-press-books.de/html/SpringerBriefs_PSP_SirPeterHall.htm.

ISSN 2194-3125 ISSN 2194-3133 (electronic)
SpringerBriefs on Pioneers in Science and Practice
ISBN 978-3-319-28054-7 ISBN 978-3-319-28056-1 (eBook)
DOI 10.1007/978-3-319-28056-1

Library of Congress Control Number: 2015958551

Copyediting: PD Dr. Hans Günter Brauch, AFES-PRESS e.V., Mosbach, Germany

Printed on acid-free paper

This Springer imprint is published by SpringerNature
The registered company is Springer International Publishing AG Switzerland

This book is dedicated to the life and work of Sir Peter Hall, a true Pioneer in Science, who was an intellectual giant, city visionary and polymath of international renown with deep roots in both geography and planning.

Whilst best known by the public as a spatial planner for most of his career, he retained a deep interest in and enthusiasm for urban and transport geography.

The International Geographical Union's (IGU) Commission on Transport and Geography and the IGU Urban Commission (Urban Challenges in a complex world) decided to jointly recognise Peter Hall's immense contribution to the understanding of city functions, planning, structures, regeneration and futures by inviting a series of expert researchers, including several of his co-researchers, to write and reflect on Peter's key research themes.

Contents

List of Figures

The above listed figures have been used with copyright permissions from the sources mentioned in the figure captions.

List of Boxes

Chapter 1
Introduction: Professor Sir Peter Hall, Pioneer in Regional Planning, Transport and Urban Geography

Richard D. Knowles and Céline Rozenblat

1.1 Life and Academic Career of Peter Hall

Appointed as Professor of Geography at the University of Reading in 1968 at the early age of 36, Peter developed a unique international dimension to his understanding of cities (Box 1.1). He soon recognised the dynamism of Asian cities through association with the University of Hong Kong in the mid 1970s and of American cities as Visiting Professor in Planning at University of California, Berkeley in 1974. For nine years from 1980 he became a transatlantic commuter simultaneously holding professorships in the UK in Geography at Reading University, and in the USA in Urban Planning and Regional Studies at the University of California, Berkeley, continuing the latter role until 1992, before becoming Professor of Planning at University College London in 1992.

Box 1.1: Sir Peter Geoffrey Hall: Personal Data, Education and Professional Positions

Personal Data
Born March 19, 1932; London, England.
Died July 30, 2014
Married, no children.
Nationality: British

Richard D. Knowles, Professor, University of Salford, UK, Emeritus Professor of Transport Geography; r.d.knowles@salford.ac.uk.

Celine Rozenblat, Professor, University of Lausanne, Switzerland, Associate Professor of Geography; Celine.rozenblat@unil.ch.

R.D. Knowles and C. Rozenblat (eds.), *Sir Peter Hall: Pioneer in Regional Planning, Transport and Urban Geography*, SpringerBriefs on Pioneers in Science and Practice 52, DOI 10.1007/978-3-319-28056-1_1

Education

1943–50	Blackpool Grammar School
1950–56	University of Cambridge (St. Catharine's College)
1952	Geographical Tripos, Part 1: Class 1 and Philip Lake Prize
1953	B.A., Geographical Tripos, Part 2: Class 1
1957	M.A., University of Cambridge (St. Catharine's College)
1959	Ph.D., University of Cambridge (St. Catharine's College). Thesis entitled: The Location of Industry in London: 1851–1939

Professional Positions

1957–60	Assistant Lecturer, Birkbeck College, University of London
1960–65	Lecturer, Birkbeck College, University of London *Main teaching responsibilities: Regional Geography of Germany; Historical Geography of the British Isles; Applied Geography*
1966–67	Reader in Geography with special reference to Regional Planning, London School of Economics and Political Science, University of London *Main teaching responsibilities: Applied Geography; M.Sc. Regional and Urban Planning Studies*
1968–89	Professor of Geography (Head of Department, 1968–80), University of Reading (Emeritus, 1989) *Main teaching responsibilities: Postgraduate: M.Phil. Environmental Planning; Supervision of Ph.D. students. Undergraduate; Basic Historical Geography; Basic Population and Urban Geography; Basic Contemporary Problems; Cities; Development*
1971–77 and 1983–6	Chairman, School of Planning Studies, University of Reading *Main administrative responsibilities: Course Coordination; Academic Development; Faculty Liaison; Publicity*
1974	Visiting Professor, Department of City and Regional Planning, College of Environmental Design, University of California, Berkeley
1975–78	Dean, Faculty of Urban and Regional Studies, University of Reading

1980–92	Professor, Department of City and Regional Planning, University of California at Berkeley (Emeritus, 1993) *Main teaching responsibilities: Graduate Courses: The Planning Process; Metropolitan Planning in the Developing World; Research Seminar*
1980–8	Associate Director, Institute of Urban and Regional Development, University of California at Berkeley *Main administrative responsibilities: Faculty Liaison; Grants and Contracts Development*
1983–6	Member of Board and of Management Committee, Joint Centre for Land Development Studies, University of Reading/College of Estate Management *Main administrative responsibilities: oversight of research programme; research development; financial management*
1989–92	Director, Institute of Urban and Regional Development, University of California at Berkeley *Main administrative responsibilities: Overall control of budget and research program; overall management of the day-to-day work of the Institute; coordination of the Institute's seminar and publications programs*
1992–2005	Bartlett Professor of Planning, University College London *Main administrative responsibilities: Direction of research*
2005–14	Bartlett Professor of Planning and Regeneration, University College London *Main teaching responsibility: M.Sc. course in Urban Regeneration*

Peter Hall's academic prowess first came to public notice with publication in 1962 of his Cambridge University Doctoral Thesis in Geography on *The Industries of London since 1861* (Hall 1962), an incisive analysis in urban economic and historical geography. Peter's sought to understand how cities function and his enthusiasm for urban planning, interests in transport and new technologies, his global perspective, and a willingness to embrace the future shone through, starting with his books on *London 2000* (Hall 1963) and his far-seeing *The World Cities* (Hall 1966), which established the concept of world city regions 30 years before it became mainstream. Whilst advocating urban regeneration and post-industrial urban renaissance, Peter was never hesitant about highlighting planning failures and his book *Great Planning Disasters* (Hall 1980) predated more recent critiques by Flyvbjerg, Priemus and others by more than 20 years (Flyvbjerg et al. 2004; Flyvbjerg 2014; Priemus et al. 2008; Priemus/van Wee 2013).

Peter single authored 13 books, many best sellers with second or more editions and some also translated into other languages (Hall 1962, 1963, 1966, 1970, 1974, 1977, 1980, 1988, 1989, 1998, 1999, 2007, 2014), and co-authored or edited 38 books, contributed to more than 160 books, published about 200 articles and research journal papers (see Chap. 8 for details) and also produced hundreds of other papers and conference presentations and numerous short contributions to the journals New Society, The Planner and Town and Country Planning. He was also closely involved with the editing of seven journals: Built Environment and Regional Studies as Editor, and Environment & Planning C, Futures, New Society, Planning Perspectives and Journal of Transport Geography as Editorial Board member.

Peter Hall's breadth of intellectual curiosity and endeavour was enormous, reflecting his roots in urban, economic, transport and historical geography, deep interest in urban and regional planning, and continuing fascination with London. Tewdwr-Jones et al. (2014) classified Peter's prodigious research output into five distinct categories and nineteen themes:

1. History of Cities and Planning. This encompassed Urban and Regional Planning, Great Planning Disasters, Cities of Tomorrow, and Cities in Civilization;
2. London's Growth and Development. This comprised London's Economy, London's future in 2000 and 2001 from a 1960s and 1980s perspective, and London as a Working Capital;
3. Spatial Planning. This included Regional Planning, the Containment of Urban England, Sociable Cities, Regeneration—the Inner City in Context, and 'Non-Plan' Enterprise Zones;
4. Connectivity and Mobility. This encompassed the Information Age and Technological change, Technopoles of the World, and Transport and Planning;
5. Globalized Urbanization. This comprised The World Cities, The Polycentric Vision, Europe 2000 and Urban Futures.

Peter Hall's hallmark as a polymath and intellectual giant is that he was able to research and publish in such a diverse range of thematic areas for over 50 years. But his great genius was his ability to link all these five aspects in a complementary way to build a holistic comprehension of the dynamic of cities (and mobility as a part of this dynamic) in the late 20th century and the beginning of the new millennium. He was able to create original visions of urban society development with precocity and clear-sightedness, helping planners and decision makers to identify the crucial tendencies and to evaluate the limits of possible actions.

Through his numerous publications, Peter sought to engage with planners, politicians and the wider public beyond the narrow confines of the academic world. This was recognised and rewarded by both Conservative and Labour UK government appointments including as a member of the South East Regional Economic Planning Council 1966–1979, Special Adviser on Strategic Planning 1991–1994, member of London Development Agency's Thames Gateway International Design Committee, Lord Rogers' Urban Task Force from 1998–1999, the Planning Research Network from 2003, the 2006 Barker Review of the planning system and the Eco-Towns Challenge Panel 2008, and Chair of Blackpool Urban Regeneration Company 2005–2008 and appointed to undertake an *Independent Review on Better Rail Stations* (Green/Hall 2009). He was also appointed to many official committees in the UK, and in Europe, Germany, Singapore, South Australia and Stockholm (Box 1.2). Peter also published 30 planning consultancy reports for organisations in the UK, Spain, Malta, Mexico, South Australia and Stockholm.

Box 1.2: Peter Hall: Other Professional Responsibilities

Membership of Government, Official, and International Committees:

American Academy of Arts and Sciences and Aspen Institute: Committee on Future of the Automobile (1990–92)
Built Environment journal—Joint Editor (1980–)
Centre for Environmental Studies Governor (1975–80)
  Member, Research Committee (1975–80)
City of Stockholm
  Adviser, Economic Aspects of the "Dennis Package", 1994
Commission of the European Communities
  Chairman, Information Committee (1977–80);
  Member, Study Group on New Tendencies of Socioeconomic Development (1975–77)
  Member, President Delors' Carrefour on Urban Development (1993–5)
Communities and Local Government
  Member, Eco-Towns Challenge Group (2008)
Department of the Environment
  Member, Planning and Transport Advisory Council (1971–73)
  Member, Environmental Board (1975–79)
  Chairman, Subcommittee on Environmental Education (1976–78)
  Special Adviser to Secretary of State on Strategic Land Use Planning (1991–4)
Department of the Environment, Transport and the Regions
  Member, Urban Task Force (1998–)
Department of Transport
  Member, Advisory Committee on Trunk Road Assessment (1977–79)
  Member, Standing Advisory Committee on Trunk Road Assessment (1979–80)
Federal Republic of Germany
  Moderator, *Weltkommission* on The City in the 21st Century (1998–2000)
Government of South Australia
  Adviser on Strategic Review for Metropolitan Adelaide, 1990–2

H.M. Treasury, Barker Review of Land Use Planning
  Member, Expert Advisory Committee, 2006
House of Commons
  Organized Special Symposium for Select Committee on Transport, Berkeley (1988)
European Commission, Interreg IIIB Programme
  Director, *POLYNET: Sustainable Management of European Polycentric Mega-City Regions*
London Development Agency
  Member, Thames Gateway International Design Committee
Ministry of Housing and Local Government
  Member, Research Advisory Group (1965–66)
Ministry of Transport (Department of the Environment)
  Member, Urban Motorways Committee (1969–72)
  Member, Bus Demonstration Projects Committee (1969–73)
Nature Conservancy Council
  Member (1966–72)
Office of the Deputy Prime Minister
  Member Planning Research Network, 2003–
ReBlackpool (Blackpool Urban Regeneration Company)
  Chair, 2005–8
Social Science Research Council
  Member (1975–80)
  Chairman, Planning Committee (1975–80)
  Joint Chairman, Planning and Human Geography Committee (1975–80)
  Member, Research Initiatives Board (1975–80)
  Chairman, Inner Cities Working Party (1978–80)
South East Regional Economic Planning Council
  Member, Council (1966–79)
  Chairman, Research Group (1969–79)
  Member, Planning and Environment Committee (1972–79)
  Chairman, Structure Plans Monitoring Group (1975–79)
South East Region Passenger Transport Coordinating Committee
  Chairman (1967–69)
Transport and Road Research Laboratory
  Member, Research Committee on Road Traffic (1966–72)
  Member, Advisory Committee on Transport (1973–)

Peter Hall was awarded numerous national and international honours and prizes including the Royal Geographical Society's Founder's Medal, George Stephenson Medal by the Institution of Civil Engineers, London, the Ebenezer Howard Memorial Medal by the Town and Country Planning Association, Prix Vautrin Lud ("Nobel de Geographie"), Royal Town Planning Institute's Gold Medal, the Balzan International

Prize, the UIA (International Union of Architects) Sir Patrick Abercrombie Prize, as well as Honorary Doctorate Degrees by 14 universities in Britain, Sweden and Canada (Box 1.3). Peter was knighted in 1998 to become Sir Peter Hall.

Box 1.3: Sir Peter Hall: Honours and Prizes

1968	Gill Memorial Prize, Royal Geographical Society
1979	Adolph Bentinck Prize
1983	Fellowship of the British Academy
1988	Founder's Medal, Royal Geographical Society
	Honorary Fellowship, St Catharine's College, Cambridge
1989	Member of the Academia Europaea
	Degree of Doctor of Social Science *Honoris Causa*, University of Birmingham
1991	George Stephenson Medal, Institution of Civil Engineers, London
1992	Degree of Doctor of Philosophy *Honoris Causa*, University of Lund, Sweden
1995	Corresponding Member, Österreichisches Wissenschaftliche Akademie
	Degree of Doctor of Letters *Honoris Causa*, University of Sheffield
	Degree of Doctor of Letters *Honoris Causa*, University of Newcastle upon Tyne
1996	Degree of Doctor of Engineering *Honoris Causa*, Technical University of Nova Scotia
1997	Degree of Doctor of Arts *Honoris Causa*, Oxford Brookes University
1998	Created Knight Bachelor
1999	Ebenezer Howard Memorial Medal, Town and Country Planning Association
1999	Degree of Doctor of Law *Honoris Causa*, University of Reading
2000	Degree of Doctor of Science *Honoris Causa*, University of the West of England
2001	Degree of Doctor of Laws *Honoris Causa*, University of Manchester
	Prix Vautrin Lud ("Nobel de Géographie")
2002	Degree of Doctor of Letters *Honoris Causa*, Heriot Watt University, Edinburgh
	Degree of Doctor of Letters *Honoris Causa*, London Guildhall University
2003	Named by Her Majesty Queen Elizabeth II as a "Pioneer in the Life of the Nation"
	RTPI Gold Medal
2004	Degree of Doctor of Social Sciences *Honoris Causa*, Queen Mary University of London
	Degree of Doctor of Technology *Honoris Causa*, University of Greenwich
	Honorary Professor, Tongji University, Shanghai, China

2005	Deputy Prime Minister's Lifetime Achievement Award, Urban Summit, Manchester
	Degree of Doctor of Science *Honoris Causa*, Loughborough University
	Balzan International Prize
2006	Elected President of the Regional Studies Association
2008	UIA (International Union of Architects) Sir Patrick Abercrombie Prize
2014	(posthumous) Alan Hay Award for significant contributions to Transport Geography, Transport Geography Research Group, Royal Geographical Society (with the Institute of British Geographers)

Peter had a lifelong interest in transport that developed from his fascination with London's Underground train network as a young child growing up in pre-war Hampstead. He recognised that transport played a critical role in the development of cities through a process that Knowles (2006) called 'transport shaping space'. He returned repeatedly to this relationship and strongly advocated the building of London's M25 orbital motorway, an Urban Development Corporation for London's disused docklands underpinned by new light rail and underground lines, Stansted as London's third airport, the Channel Tunnel between Britain and France and the Channel Tunnel high speed rail link's associated Thames Gateway, London Crossrail, the High Speed 2 rail route between London and Northern England, and better transport links between England's northern cities.

Peter was a founder member of the Journal of Transport Geography's International Editorial Board in 1993 and served until 2008, continuing to referee papers thereafter with characteristic but critical enthusiasm. He also published several papers in the Journal with his doctoral research students (Titheridge/Hall 2006; Chen/Hall 2011, 2012). Peter also supported the Royal Geographical Society's Transport Geography Research Group (TGRG) and in August 2014 was posthumously awarded TGRG's 2014 'Hay Award' for significant contributions to Transport Geography.

In urban and regional geography he was also very active. In 1983, he co-published *Changing Development Hierarchies in the Development Process: An International Comparison* for the International Geographical Union (IGU). He actively participated in numerous regional, national and international appraisals, especially for London, for the European Union and for several European countries (see Chap. 8). In the academic field, beside his numerous national responsibilities, he was elected President of the Regional Studies Association in 2006.

1.2 Structure of the Book

Following this Introduction, the book is structured into six chapters to reflect Peter Hall's main research themes, all of which relate to the spatial organization of human society.

In Chap. 2, 'A polymath in city studies', Peter Taylor explores Peter Hall's status as an inter-disciplinary polymath and five different aspects of his lifelong fascination with cities: geography and planning, cities and state, times and spaces, town/country and city/region, and London and globalization.

In Chap. 3, 'Location and Innovation', Jonathan Reades examines how Peter's deep understanding of firm location and technological innovation informed his views of urban regeneration and regional development. Peter's research emphasised the importance of history, chance, clusters, infrastructure and the role of the state.

In Chap. 4, 'Transport and place-making—a long view', Chia-Lin Chen explores Peter's abiding interest in transport and place making, and his fascination with transport as a maker and breaker of cities. Peter shaped his vision through the period where modern London was created and Britain was rebuilt, the transformative period of the great rail revival and the return to the city, and strategic planning for polycentric cities and spatial rebalancing.

In Chap. 5, 'The strategic planning protagonist: Unveiling the global mega-city region', Kathy Pain examines Peter's work as a regional planning protagonist who foresaw the emergence of global mega-city regions, and focussed on governance and planning challenges.

In Chap. 6, 'Creative destruction, long waves and the age of the smart city', Michael Batty explores Peter's research on the impact of technology on cities and regions, long waves of technological innovation, the creative city, the age of the 'smart city' and the technological future.

In Chap. 7, 'The visionary of World and European cities', Celine Rozenblat and Dan O'Donoghue consider reasons for Peter's status as a visionary of cities and assess his record of research into universal urban processes and the uniqueness of cities, urban transformations and technological change, stages in urban development, and a multi-scale approach to the study of cities.

References

Batty M. (2014) Obituary—Peter Hall 1932–2014. *Environment and Planning A*, 46, 2263–2267

Chen C-L. and Hall P. (2011) The impacts of high-speed trains on British economic geography: a study of the UK's InterCity 125/225 and its effects. *Journal of Transport Geography*, 19, 689–704

Chen C-L. and Hall P. (2012) The wider spatial-economic impacts of high-speed trains: a comparative case study of Manchester and Lille sub-regions. *Journal of Transport Geography*, 24, 89–110

Flyvbjerg B. (ed.) (2014) *Megaproject Planning and Management: Essential Readings Volumes 1 & 2*. Cheltenham UK, Edward Elgar

Flyvbjerg B., Bruzelius N., and Rothengatter W. (2003) *Megaprojects and Risk: an anatomy of ambition*. Cambridge, Cambridge University Press

Green C. and Hall P. (2009) *Better Rail Stations: An Independent Review*. for Lord Adonis, Secretary of State for Transport. London, Queen's Printer and Controller of Her Majesty's Stationery.

Hall P. (1962) *The Industries of London*. London, Hutchinson

Hall P. (1963) *London 2000*. London, Faber [Italian edition 1965; 2nd edition 1969]

Hall P. (1966) *The World Cities*. London, Weidenfeld & Nicolson [2nd edition 1977; Chinese edition 1983; 3rd edition 1984]

Hall P. (1970) *Theory and Practice of Regional Planning*. London, Pemberton

Hall P. (1974) *Urban and Regional Planning*. Harmondsworth, Penguin & Newton Abbot, David and Charles [2nd edition 1983. London, Penguin; 3rd edition 1992. London, Routledge; Chinese edition 1996. Taipei, Chu Liu Book Company; 4th edition 2002 London, Routledge; Chinese Edition 2008. Beijing: China Architecture and Building Press; 5th edition with Tewdwr-Jones M. 2011, London, Routledge]

Hall P. (1977) *Europe 2000*. London, Duckworth [Chinese edition 1982]

Hall P. (1980) *Great Planning Disasters*. London, Weidenfeld [paperback edition 1981. London, Penguin; American edition 1983, Berkeley & Los Angeles, University of California Press]

Hall P. (1988) *Cities of Tomorrow: An Intellectual History of Urban Planning and Design in the Twentieth Century*. Oxford, Blackwell [Spanish edition 1996. Barcelona, Ediciones del Serbal; 3rd edition 2002 Oxford, Blackwell]

Hall P. (1989) *London 2001*. London, Unwin Hyman

Hall P. (1998) *Cities in Civilization: Culture, Technology and Urban Order*. London, Weidenfeld and Nicolson

Hall P. (1999) *Sustainable Cities or Town Cramming?* London, Town and Country Planning Association

Hall P. (2007) *London Voices London Lives: Tales from a Working Capital*. Bristol, Policy Press

Hall P. (2014) *Good Cities, Better Lives: How Europe Discovered the Lost Art of Urbanism*. London, Routledge

Knowles R.D. (2006) Transport shaping space: differential collapse in time-space. *Journal of Transport Geography*, 14, 407–425

Priemus H. and van Wee B. (2013) *International Handbook on Mega-Projects*. Cheltenham, Edward Elgar

Priemus H., Flyvbjerg B. and van Wee B. (2008) *Decision-making on Mega-projects: Cost-benefit analysis, planning and innovation*. Cheltenham, UK, Edward Elgar

Tewdwr-Jones M., Phelps N.A. and Freestone R. (eds.) (2014) *The Planning Imagination: Peter Hall and the Study of Urban and Regional Planning*. London, Routledge

Titheridge H. and Hall P. (2006) Changing Travel to Work Patterns in South East England. *Journal of Transport Geography*, 14, 60–75

Chapter 2
A Polymath in City Studies

Peter J. Taylor

2.1 On Being a Polymath

I am using the word polymath to describe the remarkable range of Peter Hall's scholarship as a means of avoiding popular terminology that promotes disciplinary thinking through its inter- or multi- or cross-variations. The concept of polymath long precedes the disciplining of knowledges, which makes it so appropriate as a descriptor of Peter's oeuvre. An obvious reason for this failure in intellectually disciplining Hall is his background in Geography wherein strict disciplinary thinking does not make sense. But I suggest a more important influence: Peter's fascination with cities that has always been at the heart of his work, largely explicitly, otherwise implicitly. Researching cities straddles social science disciplines, fits into none but requires all.

In terms of disciplinary thinking Peter is generally associated with three, Planning and History being added to Geography (Phelps et al. 2014). Such 'tri-disciplinarity' is rare but still it hardly does justice to Peter's studies. Being interested in cities meant that economic processes have to have an important role in Peter's research and sure enough agglomeration mechanisms and corporate structures feature prominently in his work. So we might add Economics and Business Studies to the Hall intellectual mix. We can make exactly the same argument for Sociology—cities are profoundly 'social/cultural'. Furthermore the subjects Peter tackled had weighty political implications which he always tackled head on: add Political Science. On a personal note, I encountered the latter side of Peter's thinking when I was editing *Political Geography Quarterly*. On launch, we invited contributions on the nature of Political Geography as a means of charting new directions. Amongst many responses from well-known political geographers there

Peter J. Taylor, Professor, Northumbria University, Newcastle upon Tyne, UK, Professor of Human Geography; crogfam@yahoo.com.

R.D. Knowles and C. Rozenblat (eds.), *Sir Peter Hall: Pioneer in Regional Planning, Transport and Urban Geography*, SpringerBriefs on Pioneers in Science and Practice 52, DOI 10.1007/978-3-319-28056-1_2

was an offering from Peter (Hall 1982a). It was a surprise to me that he would appear in the first issue of the journal since I had thought of him at that time as a 'planner-geographer', Peter suggested a new political economy frame for our studies. This discussion has already reached seven disciplines without any mention of Transport Studies, possibly the subject he cherished most.

The point I am making is that Peter was not a sort of brilliant 'octo-disciplinarian' simply because he did not take disciplines seriously in his work. His was a very practical slant on the way he contributed to our knowledge of cities, most obviously in Planning. Taking the latter out of the argument for the moment, not for him the conventional nod to long-dead 'founding fathers' providing guidance for defining current research, rather he used ideas and concepts from cognate disciplines for the specific purposes of his own understandings. From a disciplinary point of view these past academic heroes provide intellectual anchors in socializing (i.e. disciplining) new entrants to a discipline. This is where the benefit of Peter's academic origins being in Geography is so relevant since its founding fathers are largely of 'academic' interest in the derogatory sense (i.e. how not to do Geography). Interestingly Peter did edit a translation of a seminal book by an Economics founding father, von Thünen, but for a Geography audience (Hall 1966a)! But of course, in a curious twist, Peter contributed to turning Planning from a practice into a discipline by giving this activity the requisite founding fathers to take it beyond professionalization—'architecture writ large'—and making it a respectable social science (Hall 1974),on which more below. In this case the practical side of Peter's thinking coincided with a need to discipline, which is what he provided using critical historical narrative. It is this part of Hall's legacy that is particularly celebrated in the volume *The Planning Imagination* produced to mark his eightieth birthday (Tewdwr-Jones et al. 2014). Ron Johnston also assesses Peter as being foremost a planner, more than all the disciplines mentioned above (Taylor 2014: 226). But I will treat him as a polymath of city studies.

There is an obvious danger that Peter's eclectic mix of scholarship results in superficial thinking, after all it is the purpose of disciplines to insist on 'depth' of understanding. This suspicion might be enhanced by Peter's journalism, a plethora of short opinion pieces penned throughout his career. As a young lecturer I read Peter every week in *New Society*. In this work Peter expressed his practical side (and more) in a very up-to-date (and entertaining) manner proving the worth of such interventions. If you had to pick one word to encompass Hall's writings, curiosity and passions I think the term 'Tomorrow' is especially suited. It undergirds both his journalism and his supposedly more scholarly contributions. The former is about the present but looking forward and it is commonplace to argue that History is as much about the future as the past (which I would extend to all social science research) and this is clearly seen in Peter's research. 'Tomorrow' can be treated as the hinge at the intersection of his journalism and his research and I will deploy this intersection imagery further to show Peter as polymath was not merely a well-read collector of a vast array of ideas but that he deployed them interactively to better understand cities in undisciplined depth.

2.2 Five Intersections

Choosing scholarly intersections from an oeuvre built over half a century inevitably must be selective and reflect personal bias: there is no way a comprehensive record of the innate effervescence in Peter's work can be provided here. I have selected five intersections that I think reflect prescience and also enduring relevance. Readers are invited to elaborate on my short vignettes and think of further intersections. I have ordered my intersections from general issues to more specific pairings.

2.2.1 *Geography and Planning*

When I was an undergraduate in the mid-1960s the relationship between Geography and Planning was seen as being quite straightforward: the latter was a recognized applied version of the former.[1] It was a local government profession dealing with land use that was a natural career choice for many geography graduates (including me). It was attractive because it was about making better (local) worlds, but it was also notorious for being rather mechanistic and therefore unexciting. Peter did not take this career route; he did a Ph.D. in historical geography but not following the latter's traditional predilection for the rural (Bassin/Berdoulay 2004: 65–72), his focus was industrial London.

Geography was in flux in Hall's early career with changes emanating from the USA but with their synthesis produced in the UK by Haggett (1965). Hall was on the fringes of this transformation; he made what may have been an unintentional contribution through making von Thünen's land use model available in English as mentioned previously. The theoretical basis of the Geography transition consisted of borrowing a trio of location models from Economics with von Thünen (via Hall) contributing the primary sector space, Weber the secondary sector space and Christaller/Lösch the tertiary sector space. Interestingly it was the von Thünen land use model that proved to be more relevant for Planning through urban land use modeling. However in Geography it was the Christaller model that was by far the most influential—for Bunge (1966) it qualified Geography as a science, no less. But these abstract takes on 'new' Geography by Haggett and especially Bunge were not to Hall's empirical and practical taste. This was not a suitable 'pure' science that was supposedly twinned with Planning as the applied science.

Instead Hall embarked on making Planning an exciting intellectual pursuit by providing it with a stimulating biography. Despite, or perhaps because of, Hall's Geography background, he provided a lineage of eminent thinkers and practitioners

[1]Hall (1974, 1988) represent the ideas discussed in this intersection. Their lasting relevance is indicated by the fact that both books are still in press through multiple editions, five for the former, and four in the latter case.

as Planning's own founding fathers from a wide range of backgrounds. These were people who engaged with city problems and city opportunities; traditional geographers were conspicuous by their absence due to their relative neglect of the urban (Wrigley 1965; Hall 2003; Taylor 2003). In this way Hall provided Planning with an intellectual grounding in social science. In the process his 'disciplinary history' provided an alternative view of the profession, shifting focus from boring bureaucratic mechanism to meaningful policy interventions. Who wouldn't want to be a planner?

This story places Peter in a key role even when the intersection of Geography and Planning was one of divergence. With Geography's neglect of cities overcome as urban geography came to dominate the human side of discipline from the 1970s, the intersection became more intimate with Hall right at the centre of the intellectual interactions. These are described in the remaining intersections.

2.2.2 *Cities and State*

Peter's practical approach to understanding cities and the state was firmly set in the reformist tradition of centre-left politics.[2] This had triumphed in creating the welfare state after 1945 of which physical planning was an integral part. In this way of thinking, society's problem could be and should be solved by state actions. In effect this nationalized a wide range of non-state radical traditions not least in Planning: this is made clear in Hall's planning history where, for instance, private-sponsored garden cities at the beginning of the twentieth century become transformed into a state programme of new towns at mid-century (Hall 1974). Such treatment of past radicalisms feeds into the suspicion that state actions on society are not the only solutions and are not inevitably benevolent.

Peter turned such suspicion into a critical assessment of state actions in three main contributions. First, he led a large-scale investigation into the effects of city containment policies instituted after 1945 to prevent 'urban sprawl'. Although the intention had been to support viable separate communities in terms of work/home balances, the results were very different. Cities were indeed contained but at the expense of increased social segregation, longer distance commuting, and rising costs of property and land. Thus for most people, there was a net material loss, while a few benefitted immensely. Second, Peter identified a set of large-scale projects instigated and implemented by the state that were seen as necessary but unobtainable through private sector funding. These large projects, such as London's third airport, were subject to planning procedures in a context where they cannot be disentangled from myriad outside effects and consequences. Dubbing them 'great planning disasters', with their inevitable cost overruns due to inherent, unpredictable uncertainties, Hall was discovering a societal complexity centred on the demands of dynamic cities.

[2]Hall et al. (1973) and Hall (1980, 1982b) are the main works underpinning this intersection.

Hall shared this reaction to the post-1945 state overreaching its capability with other 1970s critics (e.g. Adams 1970; Brooks 1974) and the issue of state profligacy remains relevant down to the present (e.g. Flyvbjerg et al. 2003; Shepard 2015). Third, Hall's appreciation that planning purposes and planning outcomes could significantly diverge culminated in his controversial promotion of 'non-plan'. This was an attack on top-down bureaucratic obstacles to local economic development. Noticing that it was Pacific Asian cities without national development plans that were experiencing dynamic growth rather than so-called Third World 'developing states' with their surfeit of such plans, Peter suggested that the persistence of urban poverty in British cities might be better tackled through local withdrawal from planning restrictions rather than inventing more area-based anti-poverty policy 'solutions'. Coming from a leading authority on planning this was seen by many as astonishing, even more so when taken up by a right-wing government as 'enterprise zones' and thereafter seen as an integral part of the political movement to neo-liberalism. But you do not have to be a neo-liberal to understand that state intervention is not the automatic answer to all social ills.

This intersection between complex cities and state policies—Scott (1998) has subsequently shown how states inherently simplify complex reality—derives from Peter's practical and empirical imperative: what works? It places him in a long radical tradition that is not state-centric.

2.2.3 *Times and Spaces*

Peter's initial Ph.D. research in historical geography and his subsequent focus on planning studies are both fundamentally about relations between time/period and space/place.[3] Over a wide range of publications, these dimensional framings, treated as social constructs, are deployed by Hall in a variety of resourceful ways.

I have already suggested 'Tomorrow' as Hall's single word descriptor and it is noteworthy that this signal to the future is used in the title of one of his historical books. Although I began with History as one of his three most obvious disciplines, Peter's temporal studies are as much about the future as about the past. Contemporary concepts of how cities work are applied to cities both across historical periods and through scenarios for the future. But the imaginative treatment of temporal structures comes with his contributions to the work on long waves/cycles that became popular in social science from the 1980s. Often used as changing contexts in broad overviews, Peter's emphasized their specificities as 'carrier waves' of innovations. This directly links to spatial agglomerations in cities as the loci of innovations. These are special places of economic and cultural change with distinctive characteristics—latterly identified as technopoles for viewing or

[3]This intersection covers a prolific amount of Peter's work represented by Hall (1988, 1998), Hall/Preston (1988), Castells/Hall (1994).

making the future. But this is not simply a case of exceptional spaces of places; these are entwined in ever changing spaces of flows. Here the intersection is with Peter's lifelong interest in transport, such as modal travel models in cities, and later with aspects of the communications revolution restructuring cities and their inter-relations. The time-space frames of innovation and technology both enabling as expected, and shifting in unpredictable ways, were plainly one of Peter's intellectual passions.

Clearly Hall's later collaborations with Manuel Castells are important in this intersection but there is a practical and powerful methodology that is very much Peter's. This is the thought pattern that takes a general process within which specific mechanisms are revealed: each carrier wave is a unique period within a general temporal repetition; each agglomeration is a unique place within a general spatial repetition.

2.2.4 Town/Country and City/Region

What is the spatial setting for human organization?[4] Should this setting provide the areal basis for planning activities? Answers to these questions in post-1945 Britain were answered in the reverse order. This produced a political-administrative framing for land use planning as required by the state: local governments became 'planning authorities' in law. In Peter's work this is where concern for practice and theory come together.[5]

The idea for the containment of British cities referred to earlier derives from late Victorian social reform movements that treated industrial cities as severely problematic places in need of solution. Rural/urban contrasts were embedded in local government organization and this was transferred over to statutory planning as 'Town and Country Planning'. The imagery implicit in this terminology is not one of large industrial cities but rather of small self-contained communities like garden cities.[6] But modern society is not organized through myriad small separate communities; the latter are functionally integrated into much larger city-based units. This was understood in the USA where 'City Planning' was recognized and Standard Metropolitan Statistical Areas were defined in the 1960 census by combining counties using commuting data. This idea of functional urban regions has subsequently been copied across the world including Britain. Thus Hall's terminology for his planning was functional and not divisive: 'Urban and Regional Planning'. Although regional planning in Britain was originally conceived in terms

[4]Peter's work in this intersection is specifically represented by Hall/Pain (2006).

[5]Peter was a central figure in both the Town and Country Planning Association and the Regional Studies Association through his career.

[6]The Garden City Association was formed in 1899, became the Garden City and Town Planning Association and then the Town and Country Planning Association; the statutory basis of UK planning was set out in the Town and Country Planning Act of 1947.

of large 'standard regions', with Peter active for the South East England region, only the latter had a city-region identity as London's region. But ultimately such regional planning fell by the wayside due to lack of political will, formal politics and government remained primarily organized at national and local levels. However with the rise of economic globalization articulated through world cities, city regions have made a return as global city regions (Scott 2001). With London as a classic case, Peter has been at the forefront of this movement with his POLYNET project describing and analyzing seven such regions in North West Europe. This work is notable for recognition of the innate spatial complexity of these emerging spatial structures as multi-nodal mega-city regions.

Still, despite their globality, they are defined within states so that they remain in-between national and local policy; theory with limited practice. But they are central to a different intersection in Hall's oeuvre.

2.2.5 London and Globalization

One city, London, is woven into Hall's research through all the decades of his studies.[7] In these works Peter illustrates how what goes on in London has resonance far beyond the city. In this way he shows an appreciation of scalar relations; Peter wrote about world cities long before the massive literature on globalization had even begun.

Although borrowing the term 'world city' from one of his Planning founding fathers, the prescient timing of his elaboration of the idea in the mid-1960s is quite remarkable. Seven world cities were originally identified and included New York and Tokyo along with London, which were to form the 'global trio' central to the seminal works of Friedmann (1986) and Sassen (2001) two decades later. But two other quite different 'world cities' were also identified: Randstad Holland and Rhine-Ruhr, multi-nodal regions that reappear in Hall's own POLYNET research three decades later. Peter is able to locate these different spatial structures under a single umbrella term because of his functional definition. His world cities were foremost great economic centres accounting for a relative large amount of the world's business. But they were at the same time more than this in two ways. First, they encompassed a broad range of vital 'metropolitan' functions that made them stand out politically and culturally in myriad ways. This comprehensive treatment of world cities has hardly ever been matched in later voluminous studies of such cities. Second, there is an important relational aspect to his world cities (i.e. why they are 'world'). These cities are key transport centres (great ports, road and rail nodes, and with international airports) and communication centres (publishing, media, universities) all currently researched as inter-city relations. This approach

---

[7]This intersection covers a large proportion of Peter's work, here represented by Hall (1966b, 2001), and Hall/Pain (2006).

continues in POLYNET but with more emphasis on intra-regional transport and communication. In this way his ideas are revised so that even the most primate of regions are found to be multi-nodal in contemporary globalization; for instance, London's region includes Cambridge, Milton Keynes, Reading and Southampton.

Both the initial world cities study and POLYNET combine a geographical analysis of place followed by a discussion of the planning issues, thus illustrating my first intersection Geography/Planning. But the key point in this intersection is the continuity in this research thrust combined with a critical revision of theoretical concept derived from empirical evidence and practical needs.

2.3 An Optimistic Legacy

From these five intersections we find a scholar who has maintained a resonance and relevance for over half a century. In a previous discussion of Hall's work, I have likened his immense contribution to those of Jane Jacobs and Jean Gottmann (Taylor 2014). Like Peter they had a focus on cities with cognizance of political context, using a strong empirical approach casting into the past and with an eye on the future. In the 1960s each produced a seminal book on cities upon which a current literature is still building: alongside Peter's *The World Cities* (1966b) and contemporary global cities literature (Sassen 2001; Brenner/Keil 2006), there is Gottmann's (1961) *Megalopolis* and contemporary megaregions literature (Florida et al. 2008; Harrison/Hoyler 2015), and Jacobs' (1969) *Economy of Cities* and contemporary literature on agglomeration and network externalities (Glaeser 2011; Taylor 2013).

One common characteristic of these three polymaths in urban studies was their optimism for cities in a decade when cities were beginning to be seen as ominous portends for a worrying future. Not for them 'cities as problem' but rather cities were to be seen as places where society can find solutions to its ills. This optimistic attitude meant that all three recognized and promoted the practical implications of their works. However neither Gottmann nor Jacobs engaged with the policy side of urban studies with the intensity and passion that marks Hall's work, and which continued throughout his career. The decades of Hall's writings were times of immense societal change and these have been grist to his empirical and practical approach. By concentrating on his contributions to understanding the past and the future, his role—both journalistic and academic—as an inquisitive chronicler of an ever-shifting present for over half a century can easily be undervalued. When a biography of Peter appears, as surely it must, I think it is the latter that will be the added ingredient derived from assessment of his life in comparison to specialist academic assessments of his research.

References

Adams J. (1970) Westminster: the fourth London airport? *Area* 2(2), 1–9.

Bassin M. and Berdoulay V. (2004) Historical geography: locating time in the spaces of modernity. In Benko G and Strohmayer U. (eds.) *Human Geography: A History for the 21st Century*. London, Arnold.

Brenner N. and Keil R. (eds.) (2006) *The Global Cities Reader*. London, Routledge.

Brooks E. (1974) Government decision-faking. *Transactions of the Institute of British Geographers,* 63 (4), 29–40.

Bunge W. (1966) *Theoretical Geography*. Lund, Sweden, C.W.K. Gleerup.

Castells M. and Hall P. (1994) *Technopoles of the World*. London, Routledge.

Florida R., Gulden T. and Melander C. (2008) The rise of the mega-region. *Cambridge Journal of Regions, Economy and Society,* 1, 459–476.

Friedmann J. (1986) The world city hypothesis. *Development and Change,* 17, 69–83.

Flyvbjerg B., Bruzelius N. and Rothengatter W. (2003) *Megaprojects and Risk: An Anatomy of Ambition.* Cambridge, UK, Cambridge University Press.

Glaeser E. (2011) *Triumph of the City*. London, Macmillan.

Gottmann J. (1961) *Megalopolis: the Urbanized Northeastern Seaboard of the United States.* New York, Twentieth Century Fund.

Haggett P. (1965) *Locational Analysis in Human Geography*. London, Arnold.

Hall P. (ed.) (1966a) *Isolated State*: An English Translation of *De Isolierte Staat*. Oxford, Pergamon Press.

Hall P. (1966b) *The World Cities*. London, Weidenfeld.

Hall P. (1974) *Urban and Regional Planning*. Newton Abbot, UK, David and Charles.

Hall P. (1980) *Great Planning Disasters*. London, Weidenfeld.

Hall P. (1982a) The new political geography: seven years on. *Political Geography Quarterly,* 1 (1), 65–76.

Hall P. (1982b) Enterprise Zones: British origins, American adaptions. *Built Environment*, 7, 5–12.

Hall P. (1988) *Cities of Tomorrow: An Intellectual History of Urban Planning and Design in the Twentieth Century*. London, Weidenfeld and Nicolson.

Hall P. (1998) *Cities in Civilization: Culture, Innovation and Urban Order*. London, Weidenfeld and Nicolson.

Hall P. (2001) Global city-regions in the twenty-first century. In Scott A. J. (ed.) *Global City-Regions: Trends, Theory, Policy*. Oxford, Oxford University Press.

Hall P. (2003) Geographers and the urban century. In Johnston R. and Williams M. (eds.) *A Century of British Geography*. Oxford, Oxford University Press.

Hall P., and Pain K. (2006) *The Polycentric Metropolis: Learning from Mega-City regions in Europe*. London, Earthscan.

Hall P. and Preston P. (1988) *The Carrier Wave: New Information Technology and the Geography of Innovation, 1846–2003*. London, Unwin Hyman.

Hall P., Drewett R., Gracey H., and Thomas R. (1973) *The Containment of Urban England.* London, Allen and Unwin.

Harrison J. and Hoyler M. (eds.) (2015) *Megaregions: Globalization's New Urban Form?* Cheltenham, Edward Elgar.

Jacobs J. (1969) *The Economy of Cities*. New York, Vintage.

Phelps N.A., Tewdwr-Jones M., and Freestone R. (2014) Geography, History and the Planning imagination. In Tewdwr-Jones M., Phelps N. A. and Freestone R. (eds.) *The Planning Imagination: Peter Hall and the Study of Urban and Regional Planning*. London, Routledge.

Sassen S. (2001) *The Global City*. Princeton, NJ, USA, Princeton University Press.

Scott A.J. (ed.) (2001) *Global City-Regions: Trends, Theory, Policy*. Oxford, Oxford University Press.

Scott J. C. (1998) *Seeing Like a State*. New Haven, CT, USA, Yale University Press.

Shepard W. (2015) *Ghost Cities of China*. London, Zed Books.

Taylor P.J. (2003) Global, national, local. In Johnston R. and Williams M. (eds.) *A Century of British Geography*. Oxford University Press, Oxford.

Taylor P. J. (2013) *Extraordinary Cities: Millennia of Moral Syndromes, World-Systems and City/State Relations*. Edward Elgar: Cheltenham.

Taylor, P. J. (2014) The World Cities. In Tewdwr-Jones M. et al. (eds.) *The Planning Imagination: Peter Hall and the Study of Urban and Regional Planning*. Routledge, London.

Tewdwr-Jones M., Phelps N.A. and Freestone R. (eds.) (2014) *The Planning Imagination: Peter Hall and the Study of Urban and Regional Planning*. Routledge, London.

Wrigley E.A. (1965) Changes in the philosophy of Geography. In Chorley R.J. and Haggett P. (eds.) *Frontiers in Geographical Teaching*. Methuen, London.

Chapter 3
Location and Innovation

Jonathan Reades

3.1 Introduction

Peter Hall's views of urban regeneration and regional development were informed by a deep appreciation of the importance of firm location and the impact that technological change—especially in communications—were having on corporate location preferences and patterns.[1] Witness to the decline of manufacturing in Britain, to the rise of new industry in Silicon Valley, and to the resurgence of a services-based London, Hall's research tracked a wider preoccupation with innovation in the 'high tech' and 'knowledge-intensive' sectors. Although some of his writing on these topics is now more than fifty years old, Peter Hall's mastery of both the detail and the context gives his work continued relevance: he was almost unique in being able to build a 'big picture' vision from many, many books' worth of facts.

However, the breadth of Hall's writing—and the tendency for his research in one area to inform his thinking in another—presents something of a problem: it is, frankly, impossible to prise apart his views on location and innovation from those on cities and regions. Peter was, as the second chapter of this book notes, a polymath, and so his thinking on industrial location and innovation was informed as much by his love of cities, transport, and nearly-forgotten Russian economists as by his education as a geographer at Cambridge (Batty 2014). To his diverse interests Hall married a passion for what Llewelyn-Davies called 'academic journalism' (Hall 2014: 271); it is the resulting clarity of communication that gave Peter's

Jonathan Reades, Dr., Kings College London, UK, Lecturer in Quantitative Human Geography; jonathan.reades@kcl.ac.uk.

[1]I would like to acknowledge both the support of the Balzan Foundation, whose generous award to Peter Hall made possible my Ph.D. research, and the patient, thoughtful feedback of Basak Demires Ozkul, whose comments on very early drafts of this chapter have significantly improved its overall quality. Any errors or omissions remain, of course, the author's alone.

R.D. Knowles and C. Rozenblat (eds.), *Sir Peter Hall: Pioneer in Regional Planning, Transport and Urban Geography*, SpringerBriefs on Pioneers in Science and Practice 52, DOI 10.1007/978-3-319-28056-1_3

wide-ranging books impact beyond academe and which also now enable us to make sense of the many cross-cutting links between trains of thought on the decline and renaissance of cities and industries, the evolution of urban form and function, and location and innovation.

So although I had originally intended to tackle this chapter in a broadly chronological manner, charting the evolution of his writing from an early focus on traditional industrial districts, through the emergence of ICT technopoles, and ultimately to the Advanced Producer Services of the mega-city region, that approach proved unworkable: it was impossible to sensibly discuss ideas that were latent in Peter Hall's early writing without needing to skip ahead to texts where he was much more explicit in his thinking. However, the more I read—or re-read—of Peter Hall's work, the more I was struck not so much by the sea changes he had observed, as by the underlying continuity of his ideas. Fortunately, this new appreciation of his body of work presented a new way to give my own text some shape: this chapter will focus on how key themes in Hall's first book, *The Industries of London since 1861* (1962), reappear in various guises across a publication record spanning fully fifty years.

To help contextualise these themes, I will be drawing heavily on two *Apologia Pro Vita Sua* written towards the end of Peter Hall's career: the first was given to the Balzan Prize Committee in 2005, and the second appeared in *The Planning Imagination* (2014). In these apologiae Peter gave his own perspective on his contributions and I see no reason not to let him speak in his own voice wherever possible.

3.2 The Importance of History 1

Peter Hall is perhaps best known outside of the planning community for the massive work of historical scholarship that is *Cities in Civilization* (1998), but his *magnum opus* didn't spring, Athena-like from a newfound interest in two thousand years' worth of urban invention, it was the culmination of a series of publications that roped in ever-greater swathes of industrial and urban history. And the very *first* link in this chain of works was *The Industries of London since 1861* (1962), a book adapted from his geography Ph.D. thesis at Cambridge, which set out "to try to understand why industry located in London, and particularly in certain parts of London, over the previous 100 years" (Hall 2005).

Briskly setting the tone for the rest of the work, Peter Hall begins: "This book is a study in economic geography, treated historically" (Hall 1962: 9). But in what may come as a surprise for readers familiar with Hall's later work on the polycentric metropolis and used to thinking about London from a 21st Century perspective is that the book focusses primarily on manufacturing—both the 'older' sorts of clothing, furniture, printing, and the then 'newer' endeavours of general engineering, electrical engineering and vehicles—while the role of services is addressed principally in the introduction and conclusion. In fact, the book's chief purpose is to counter the received wisdom London had *become* a manufacturing centre with the onset of mass industrialisation in the Victorian era by noting the extent to which it had always *been* such a centre. *The Industries of London since 1861* therefore has as its principal interest how 'the industrial geography of London at the present time…is related to the evolutionary process which has produced it' (Hall 1962: 9). So although Peter Hall's more recent publications (e.g. Hall/Pain 2006) attached a great deal of importance to 'Advanced Producer Services', he also recognised the vital importance of less glamorous supporting players, such as hospitality and logistics, to the success of today's urban areas.

3.3 The Importance of Chance

Supporting his argument on the evolution of these industries is an extensive—nay, exhaustive—analysis of the available secondary data. In spite of the significant technical constraints of the day, the book is full of tables generated without the benefit of a computer and of maps created without access to a GIS. This approach established a pattern that Peter continued to use throughout his career: the blend of historical sources and secondary data to illustrate how a confluence of big trends and seemingly minor events which have produced specific spatio-industrial outcomes. These are then tied together in a wider synthesis from which spatial and policy implications can be abstracted.

To give you a sense of the expanse of Hall's industrial interests: *The Carrier Wave* (Hall/Preston 1988) examines the histories of the telegraph, telephone, radio, and electrical engineering since 1846; the introduction to *Cities of the 21st Century* (Hall 1991) covers in some detail the origins of faxes in the 1930s and email in the 1980s; while *The Rise of the Gunbelt* (Markusen et al. 1991) and *Western Sunrise* (Hall 1987) trace the importance of military contracting from the 1920s to the late 1980s to innovation. Having spent my early years in technology consulting and, consequently, knowing little of Peter's work prior to my foray into planning at UCL in 2006, I was rather shocked when—without notes—he delivered a detailed, and more importantly to me, insightful lecture on the computer industry.

Of course, for many of the topics he examined, there simply isn't any long-run data on which to draw. In seeking to cover off one hundred years of London's history, *The Industries of London since 1861* is forced to devote several pages (Hall 1962: 12–19) to an examination of 'Source Materials and Their Problems': classifications change, sampling methods change, and what is reported changes as well. Nonetheless, he actively set his Ph.D. students, from John Goddard to Chia-Lin Chen, collection tasks nearly as insurmountable as the one he had set himself (see comment in Hall 2005): I was tasked with finding a telecoms operator willing to share usage data on communications flows (see, for example, the results presented in Reades/Smith 2014), while Ozkul (2014) was set to work on trying to infer changes to commuting behaviour by socioeconomic class from wildly incompatible Census data.

So Peter was quite happy to engage with data, but in subsequent works he tended to draw more heavily on the full extent of his historical knowledge—buttressed by a seemingly limitless and nearly photographic memory—to provide the rich detail that could bring to life a long passage on, say, the reasons for the emergence of a German telegraph industry with a brief, potted biography of Werner Siemens (Hall/Preston 1988: 41–42). I'm half-inclined to call these insertions 'telling anecdotes', but I think that for Peter facts and data readily came together in the service of a larger argument; they are not just 'colour' to liven up a long passage, but integral elements of the outcomes he observed. Ultimately, Hall was not a 'researcher' in the sense that he conducted a great deal of primary research on novel data sets that he himself had collected; rather, he was particularly adept at drawing together disjoint findings into a whole that was somehow greater than its constituent parts (see, for example, Batty 2014: 2265); Peter himself would have been the first to admit this, writing of *The Carrier Wave* that it is "a work of scholarly compilation rather than of original research" (Hall/Preston 1988: 33).

I would also argue that the memory for detail is what gave Peter his deep appreciation of the contribution of happenstance—'time and chance' (Hall 2000: 648)—to today's industries and conglomerations. Crucially, this meant that he never lost sight of how the inexorable march of a line or bar chart could conceal many 'near misses' or, worse, 'near hits' along the way (see, for example, Hall 1997: 302–304). Peter Hall's books overflow with the details that make it clear just how accidental—the illness and unlooked-for return to Stanford of legendary dean of Electrical Engineering, Fred Terman, for instance—some of today's seemingly inevitable successes truly are. This, in a nutshell, was Peter Hall's approach to the study of industrial location and innovation: to dive deep into the specific histories and trajectories of inventions, or groups of related inventions, ranging from radios to guns, electricity to servers, and then to assemble these pieces into a broader, synthetic theory that nonetheless left room for both chance and history.

3.4 The Importance of History 2

The Industries of London since 1861 also emphasises the importance of external economies: that the industries of London are "clustered in dense industrial quarters where many small workshops could specialise and gain economies of scale" (Hall 2005). Peter Hall is here openly "borrowing theory from Marshall" (Hall 2014) on the importance of the 'in the air' quality of small, densely-packed industrial districts to long-run innovation, particularly in the face of a shift to large, vertically-integrated Fordist production methods. And he was not slow to note the 'great irony of his long career' that the Marshallian or *neo*-Marshallian district should prove so resilient in "such dynamic regions of the world as Northern Italy and California's Silicon Valley" (Hall 2005).

Hall was also not shy of noting that Marshall "invented the New Economic Geography in 1890 only for his fellow-economists to forget all about it"

(Hall 2014: 270). Peter did not, I think, have a lot of time for purely econometric reasoning; it was relevant principally insomuch as businesspeople and planners thought it was when taking decisions (Hall 1962: 169). In particular, my sense is that Peter felt that the mathematical models advanced by modern micro- and macro-economists were not very good at dealing with the dynamic forces 'that cause industries to expand or decline at different times and in different places' (Hall 1998: 292). And since it was these dynamics and the spatial impacts that these had on cities and regions that were his principal interest, equations were therefore primarily useful as evidence, not explanation: aside from the formula for the Location Quotient (Hall 1962: 16) I am hard-pressed to recall a single instance of anything statistical in any work that he sole-authored.

Peter Hall's interest in the history of the geographical and planning disciplines was consequently not purely academic, if you will, as it's rather clear the he felt that many of the 'great books' were still profoundly *relevant* as tools for thinking, given the right context (see, for example, Ozkul 2015). In the course of my own Ph.D., it was a rare doctoral supervisory meeting that didn't result in a list of at least three *more* books (not to mention the endless articles) to read. At Peter's gentle instigation I often found myself—as he once did—in the British Library Reading Room and scouring secondhand bookshops for copies of neglected classics: Weber (1909 [1969]) especially, but also Von Thünen (1826 [1966], for which Peter wrote the introduction to the English translation), Haig (1926a, b), Coase (1937), and Lösch (1954 [1973]). For Peter, theory was useful inasmuch as it enabled us to see wider patterns in the challenges facing real, unique cities and regions. Theory was therefore a useful tool for thinking more deeply about the particularities of a given place and the kinds of interventions that might be deployed, but at the end of the day it 'will not take us all the way' (Hall 2000: 643).

I think that the limitations he observed in contemporary models—and his voratious reading habits—encouraged him to look back to the foundational thinkers in the discipline, many of whom were less concerned with formal mathematical rigour than in debating causes and consequences across discplinary boundaries. In the introduction to *Cities of the 21st Century* (Hall 1991), for instance, the growth of information economy framed in terms of 'links to the growth of the highest order centres' (Hall 1991: 5) which connects his analysis to earlier Christallerian models of how systems of cities, towns, and hinterlands might operate in relation to one another. Christaller, like Hall after him, was interested in how the interactions between cities and transportation affected the availability of goods and services. Figure 3.1 shows in schematic form Christaller's well-known functional urban hierarchy, but it is less well-known that he envisioned technological change in infrastructure not only changing a market's 'orientation' (Christaller 1933 [1966]: 72–77), but also bringing about the rise and fall of competing urban areas (1933 [1966]: 99–114).

Peter Hall notes, of course, that Christaller's neat, hierarchically nested 'tree' cannot hold in an environment (Hall 1997: 312) informed by the supra-national context of the links between 'world cities' and the polycentric structures of 'mega-city regions' (Hall/Pain 2006). However, this does not mean that the concept

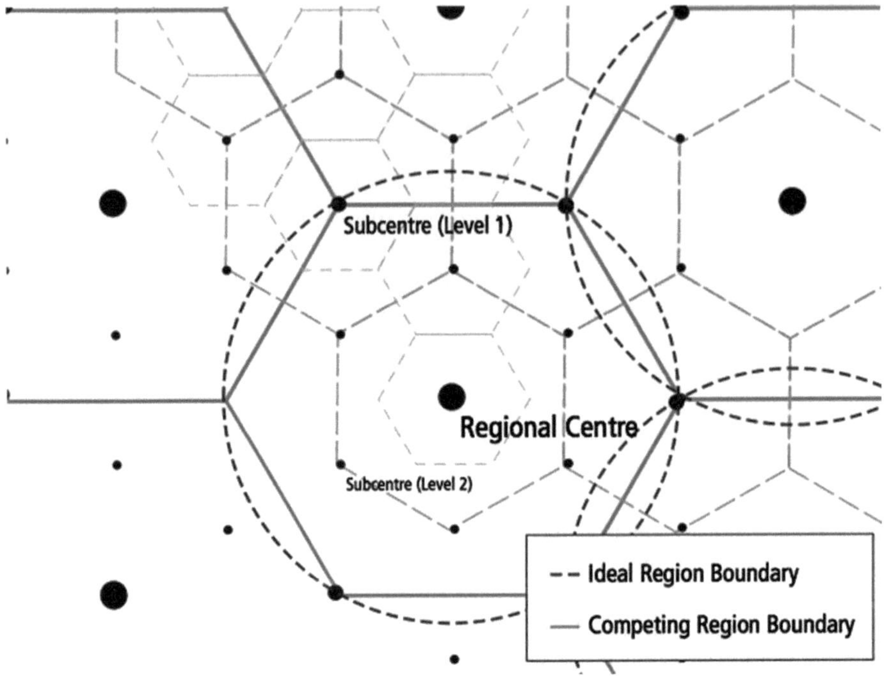

Fig. 3.1 Christaller's functional urban hierarchy. *Source* After Christaller (1933)

of 'centrality'—by which a higher order centre draws in customers to consume the complete set of goods and services that only it can provide—is obsolete; I think it's clear that for Hall this approach has relevance to our understanding of the locational patterns of many industries, especially consumer-oriented ones. Taylor et al. broadly concur, arguing that Christaller shouldn't be consigned to the 'dustbin of history' when what was missing was a complementary mirror-view of flows which properly captured the dynamics between cities in a network: "in central place theory, places make flows; in central flow theory, flows make places" (Taylor et al. 2010: 2815). But in many ways Peter Hall was there first, having already identified the need for a 'new theory of location' that would "…start with a neo-Weberian model of the informational economy, on top of which would be superimposed a Christaller-Lösch system of central places, modified to take account of changes since they wrote…" (Hall 2003: 148).

So although recent retrospectives (see, for example, Tewdwr-Jones et al. 2014) have tended to focus on Peter's obvious *penchant* for Marshall and, to a lesser extent, Christaller, I think there are other lines of thought in play here that call for our attention. In particular, I think that it is important to note the extent to which *The Industries of London since 1861* takes an unmistakably Weberian view of industrial location: Weber's analysis has sometimes been dismissed as mere 'geometrical reasoning' using the concept of the ton mile—which is the cost of shipping one tonne of a good over one mile—to find the location that minimises the total cost of moving inputs from source to factory and of moving outputs from factory to market. The argument here is that transport-cost decisions by firms help give rise to both the dense industrial districts observed by Marshall and the dispersed factories and foundries of Weber's Germany. Figure 3.2 illustrates this using Weber's 'locational triangle' for three scenarios: when the costs of moving inputs are high, when the cost of moving outputs is high, and when skilled labour (see, for example, Weber 1909 [1969]: 107–108) is integral to the production process.

Hall hews quite closely to Weber's model in discussing how the cumulative costs of materials (cloth and power, as well as specialist subcontracting), taking outputs to market, and the availability and suitability of labour (especially in terms of the depth of the market) shaped the location of the clothing industry in London (Hall 1962: 50–66). A similarly Weberian world view colours Hall's discussion of the furniture trade and *its* inputs and outputs (1962: 82–93). Over time, the spatial structures of cooperation and competition between diverse firms interact with urban form and function to dramatic effect (1991: xii).

We can trace this terminology through Hall's subsequent work, with explicit reference in *Cities of the 21st Century* to the 'Weberian triangle' of inputs and outputs as married to "...the costs of information transfer, whether in the form of packages, or electronic impulses, or the physical movement of brains and the rather weightier bodies that unfortunately have to travel with them" (Hall 1991: 3). I can find no indication beyond a comment in *Cities in Civilization* (Hall 1998: 282) that Peter further formalised his thinking here and so I will have to fill in some gaps myself, but on the surface of things we face quite a challenge in translating Weber's locational theory for use with contemporary services and digital products where the cost of transport is effectively zero. This is where a return to the Weber's *urtext*—a return to foundational works of the sort that Peter often advocated to his students—is so useful: first, because it quickly becomes obvious that it is possible to dispense with the ton-mile costs of raw data entirely if we treat it as a 'ubiquitous resource' (Weber 1909 [1969]: 51); and, second, because it also becomes clear that the

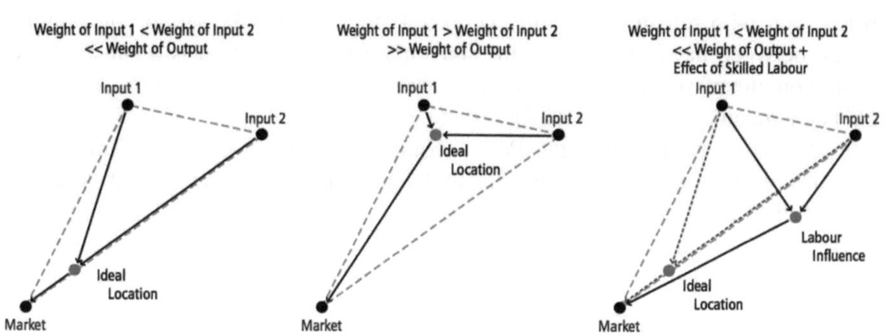

Fig. 3.2 Weber's locational triangle: 3 scenarios—high input costs, high output costs and skilled labour. *Source* After Weber (1909)

removal of one dependency merely *increases* the relative 'weight' of the remaining ones.

Though often overlooked in summaries of Weber's work, ubiquitous commodities are "so extensively available within the region that, wherever a place of consumption is located, there are either deposits of the commodity or opportunities for producing it in the vicinity" (Weber 1909 [1969]: 51). In other words, they are available in such abundance, either in absolute terms or relative to other possible sites of production, that their use by firms is both non-rivalrous and non-exclusive. Historically, the physicality of data in the form of written or printed reports—not to mention the many indirect types of information generated by, for instance, the docking of a particular boat in a port—gave a competitive advantage to traders in information who were close to transport nodes (see, for example, Vance 1970); as Hall observed: "London has an advantage as the greatest national focus of both internal and external lines of trade" (1962: 114). But digital data in its myriad forms is uniquely placeless: obviously, the geography of advanced telecoms services and hardware is far from uniform, but the barriers to shipping gigabytes' or terabytes' worth of data keep falling year-on-year. In theory, a data-dependent firm could now locate anywhere, but take away this spatial constraint and we are left with just one dependency: skilled people. Paradoxically, the range of locations available to a firm therefore hasn't widened at all since the local advantage of being 'close' to the data has been largely superseded by the global availability of labour!

I use the term 'data' above deliberately since it is manifestly *not* the same as knowledge. We could subsume knowledge under Weber's 'labour index' and make it a function of, for instance, the creative people who populate Soho's advertising agencies or East London's art studios; however, in as much as knowledge evidently *can* be transmitted between individuals and firms we clearly need to flesh out this sketch a little further. But if we are willing to place these on a continuum of mobility then we can not only cope with the differing informational demands of different industries, but we can also subsume much of the more recent literature on tacit and codified knowledge flowing from Polanyi (1967) within a wider locational framework. A less narrowly physical view of such networks makes it even easier to understand how the relative mobility of 1s and 0s, and the relative immobility of 'gifted men who refuse to migrate' (Lösch 1954 [1973]: 23), sets up an ongoing tension between the fact that "(1) the movement of information is a substitute for the movement of people; (2) it is an imperfect substitute and may be of poorer quality or quantity" (Hall 1991: 4).

In the kind of insight that Peter would have liked, Andrew Tanenbaum admonished that we should "[never] underestimate the bandwidth of a station wagon full of tapes hurtling down the highway" (Tanenbaum 1996: 91), but Peter clearly believed that the densest information storage and processing device remained the human brain. And this, in turn, leads us back to the kinds of innovation that seem to characterise both industrial districts and, in more complex ways, leading world cities.

3.5 The Importance of Clusters

These points about the imperfect substitution of data for people and of the resulting benefits from agglomeration hardly sound revolutionary today, but they were rather less taken-for-granted in the 1990s (Malecki 2014: 181). Indeed, Peter Hall made these points well in advance of the much-maligned 'death of distance' argument (Cairncross 1997),[2] but a surprising number of policymakers seem not to have grasped that this implies some serious thinking is needed around incentives: North Carolina's tax subsidies for tech firms have attracted server farms, but little in the way of employment, while San Francisco's offering to the Valley's tech firms has gone down as the 'Twitter Tax Break'. The necessity, and success, of either programme is very much open to debate. Whilst acknowledging that it was an outdated analogy even when he wrote it, Hall illustrated the informational benefits of clustering with the example of a library: the density of information embodied in a major store of knowledge such as a world-class library—or skilled set of individuals—creates a steep 'informational gradient' which effectively shuts out anyone at any meaningful distance. Leaving this library also means losing access to the critical 'unprogrammed' interactions around the water cooler or in a trendy bar which can lead to novel insights (Hall 1991: 4).

One of Peter Hall's first doctoral students, John Goddard, researched this issue in some detail (see, for instance, Goddard 1973; Goddard/Morris 1976), finding that it was the most routine functions that were most amenable to electronic substitution. Innovation, in other words, requires a large number of 'external' interactions (Goddard 1975: 50). But in *The Industries of London since 1861*, Hall had already linked the growth of clusters to 'self-generating' growth stimulated by spin-off 'swarming' (Hall 1962: 162).

In the mythos of Silicon Valley, the innovative explosion that gave us a swarm of household names arose from informal exchanges at meetings of the Homebrew Computer Club and over drinks in Walker's Wagon Wheel tavern. Coming from Britain, however, Hall noted a second dynamic in play: a way of *thinking* that he described as a 'frontier' mentality, which developed in 'cities with high net immigration, predominantly young, experimental and non-traditional, strong but informal mechanisms for exchanging information, barriers to diffusion are low, the search for novel ideas is foremost, and synergies between individuals and groups are high' (1998: 302). In other words, we need environments that are fundamentally *open* and, not coincidentally, this period saw fruitful collaborations with Ann Markusen (Hall/Markusen 1985; Hall et al. 1983; Markusen et al. 1986), Manuel Castells, and AnnaLee Saxenian. Castells himself traces the origins of *The Network Society* back to "…a series of exchanges [in the early 1980s], in Berkeley, between

[2]This is a good example of a book that few have actually read but may are happy to disparage: Cairncross' argument that cities would become important as sites for the leisure and entertainment is hardly as wide of the mark as many would imagine from the title of the book itself (see especially Cairncross 1997: Chap. 9).

Peter Hall, the late Philippe Aydalot, and myself" (Castells 1996 [2000], footnote on p. 419).

The front-row seats held by Peter and Manuel for the rise, and rise again, of Silicon Valley led to *Technopoles of the World* (Castells/Hall 1994) in which they wondered if the successes of California could be replicated elsewhere; their conclusion: "it isn't at all easy; it takes a long time; and you'd better be realistic about what you can hope to achieve" (Hall 2005). But Hall has argued, compellingly I think, that the industrialisation of research by universities and the R&D spending of governments and corporations has played a critical role in generating this 'silicon landscape'. Integrating all of these new ideas was the definitive history to be found in *Cities in Civilization* (Hall 1998) about which, though not normally considered a work on industrial location, Hall nonetheless wrote that it "was to be an extension on to a much larger canvas of the basic ideas that had been developing in all those books on innovation. Now however the theme was to extend…into artistic creativity and also…urban innovation" (Hall 2014: 279). What gives this work continued relevance is the wide-ranging view that it takes of 'creativity', placing cultural, technical, and even political innovation on an even footing, and highlighting how these underpin 'golden ages' for the cities where they unfold. In other words, innovation is not just about patents in the natural sciences and commercialisation, it is about models and processes, theatres and factories. So while Peter believed that the next wave of industrial innovation would "[come] out of scientific research and so was associated with strong universities" (Hall 2005), he nonetheless, I think, recognised the importance of the many different ways in which cities and regions could foster and cultivate individual and corporate creativity.

3.6 The Importance of Infrastructure

I noted above that it is very difficult to prise industrial location apart from Hall's many other interests, and the importance of infrastructure to the quasi-Weberian network model I have been advancing makes this issue obvious: to Peter's way of thinking, infrastructure plays a critical role in shaping the extent to which firms find locations to be attractive. I will not go into more detail here as this topic is being covered in greater depth elsewhere in this book, but the importance of infrastructure is explicit in both Weber's and Christaller's models: good roads bring places closer together, while poor ones push them apart. This is not to suggest that smaller town and cities necessarily stand to benefit from propinquity: the Christallerian view is that they may end up being bypassed altogether as jobs and consumption are hoovered up by more central places; or, as Hall noted: speed increases may reduce the importance of secondary cities (Hall 1991: 12).

What Peter's lifelong advocacy of transport—and particularly of High Speed Rail—pointed towards is the idea that infrastructure investment could 'bend the curve' (Hall 1997: 307), but only if it came in the right place, at the right time, and in the right amount. Given the spatial and economic impacts of rail developments/redevelopments such as Amsterdam's Zuidas, the 'Knowledge Quarter' at King's Cross, and France's many TGV-led projects, it is clear that transport infrastructure can have an enormous impact on industrial and urban geography. I think it is fair to say that he found the British transport planning model somewhat lacking when stood alongside the French and German ones.

Peter Hall's later works tended to focus on generic *functions*, particularly in the context of increasingly complex specialisation at the regional level (e.g. Hall/Pain 2006) and the trend towards 'concentrated deconcentration'. This approach can sometimes seem quite removed from the more mundane reality on the ground, but I think that makes it easy to overlook the fact that he was attuned to the importance of the actual premises on offer. In *The Industries of London since 1861* this can been seen in the reference to the utility of "low-rent slum premises" (Hall 1962: 179) where returns are unpredictable and flexibility is essential. These premises are integral to innovation because they lower the barriers to entry by permitting the entrepreneur to think: "I will have a shot" (1962: 167). In contrast to 'planned' spaces such as research parks or newly-produced 'artists' quarters', the Old Street area of London has been home to clusters of one sort or another since Booth and Potter (Hall 2014: 270). The endless adaptability of older industrial spaces is a crucial element in the revitalisation of the inner city economy (see, for example, Hutton 2004, 2006, 2008), and this neatly ties together Hall's interests in industry with his outputs in the realms of transport and the built environment.

3.7 The Role of the State

To the extent that government can make these sorts of spatial and economic transitions feasible policy has an important role to play in urban and regional development (Hall 1991: 13). At the time Hall wrote *The Industries of London since 1861*, the government's Board of Trade had in place a policy designed to restrict the growth of London's manufacturing industries by constraining their ability to adapt existing factories or move to larger premises (Hall 1962: 176). We could argue that this policy might not have been the right one since it merely accelerated the decline of skilled and semi-skilled manufacturing within the metro region—a process that seems to be reaching its apogee at the start of the 21st Century—but we cannot pretend that it had no effect at all.

Of course, big projects call for a big government with, at the very least, a deep wallet and Castells and Hall wrote approvingly of Japan's willingness to take a long view of investment (1994: 75). There is abundant evidence in Hall's writing that he saw a vital role for public ownership of, or at least very strong regulation to do with, major communications and transport infrastructure: in the UK, widening use of the telegraph—and resulting innovations in dependent industries—was clearly stimulated by nationalisation, which fostered interoperability and vastly expanded the market (Hall/Preston 1988: 43). In a contemporary context, decisions made by the EU about the standardisation of mobile telecommunications created a wave of

European 'winners' in the first major growth industry of the 21st Century. The subsequent fall of many first-wave consumer-facing brands, such as Nokia, is a salutary reminder of the speed with which the dynamics in play in contemporary industrial sectors can throw an industry into disequilibrium. And so even if the importance of the 'Gunbelt' (Markusen et al. 1991)—which Peter Hall considered his 'most insightful' work (Hall 2014)—is not *quite* as overwhelming as sometimes suggested (see review in Bury 1992), government policy has clearly played a crucial role in the industrialisation of research and innovation.

Translating his understanding of the military's role in regional development to the British context, Hall discerned the outlines of an emerging 'M4 motorway corridor' to the west of London in *Western Sunrise* (1987). But in the British case, the heavy hand of military spending skewed investment away from both fundamental research and civilian commercialisation (Hall/Preston 1988: 236–252) with the result that British technology firms simply could not keep pace with the explosive growth of Silicon Valley. However inadvertently, American policy established a space for 'creative excess' from which early Valley firms could benefit in the pre-commercialisation phase, while British spending constraints and the absence of secondary demand simply led to wastefulness; too little money of the wrong sort (Hall/Preston 1988: 279). Without the competitive pressures generated by a mix of government and private R&D (Castells/Hall 1994: 82), dependency on a small number of state champions causes no small number of problems for industry and government (Hall 1997: 306).

Cumulatively, the evidence of this group of works—*The Carrier Wave*, *Cities of the 21st Century*, and *The Rise of the Gunbelt* (Hall/Preston 1988; Hall 1991; Markusen et al. 1991)—is that while military money and government-mandated or managed infrastructure are often necessary, innovative users or consumers on a larger scale are required before products can find their way to a mass market (Hall/Preston 1988: 281). This, again, is the frontier mentality: in the ferment enabled by the meeting of so many different minds with so many divergent world views, some cities find themselves at the vanguard of a new wave of innovation in which ideas are opportunistically cobbled together to create new industries one after another. The challenge that this poses for the state is that it is, by its nature, conservative and poorly equipped to cope with the socio-spatial impacts of periods of 'creative destruction'.

3.8 Conclusion

At the IinteR-La+b conference in Lugano (2014) Peter sought to envision the 'growth businesses' of the 21st Century, settling on 'eco-tech' for managing environmental impacts, the marriage of art and technology in new cultural-technical forms, and the birth of sectors for which the Internet is a utility like any other. I would suggest that, in the context of the themes set out in this chapter, these are industries uniquely suited to real cities, and not to manufactured technopoles or 'science cities'. It is real, messy, historical cities that: (a) face the severest ecological tests from a mix of internally-generated and externally-generated factors; (b) are most dependent on technology to manage increasingly complex and flexible patterns of use; (c) are the most conducive environment in which new sectors, calling for disparate expertise, can be incubated; and (d) retain a significant lead in terms of the breadth and depth of their communications and transport infrastructure.

This takes us a long way from the London of 1861, and some of the technologies at our disposal today would have defied even Peter's inventive imagination as a newly-minted Ph.D. in 1960. But his work displays an active appreciation of the ways in which technology and locational preferences interact with cities and infrastructures to generate complex feedback effects; a particularly nice illustration of this integrative approach can be found in *Cities of the 21st century: new technologies and spatial systems*:

> The fortunes of these world cities, and those that depend on them at lower levels in their local hierarchy, will clearly be determined by both their traditional and new roles as hubs or central nodes shaping the trunk lines of these networks… Cities themselves will become increasingly multi-centred around information-based industries, and the development of local centres will help to shorten travel generally and commuting distances in particular. And recreational facilities may be expected to be increasingly important components in (and between) these centres, as the quality of life increases in important in the hierarchy of human need (Hall 1991: xii).

In one paragraph Hall draws together world cities, firm location, the modern polycentric city, and the role of information and ICT in urban and regional development. This is, quite simply, the ongoing agenda of a good proportion of regional studies, and they are the ideas to which I find myself turning with growing appreciation as I try to make sense of my own research on business location and innovation, and its often ambiguous results. However, in spite of the distance travelled in this chapter it seems somehow natural to bring this work to its conclusion with the closing sentence of Peter's first book: "Only by the study of what has been, and what is, can we legitimately proceed together to the study of what might be" (Hall 1962: 181). I cannot say it any better than that.

References

Batty M. (2014) Obituary—Peter Hall 1932–2014. *Environment and Planning A*, 46(10), 2263–2267.

Bury E. (1992) The Rise of the Gunbelt by Ann Markusen, Peter Hall, Scott Campbell and Sabina Deitrick. *Berkeley Planning Journal*, 7, 166–173.

Cairncross F. (1997) *The death of distance: how the communications revolution will change our lives*. Cambridge, MA, USA, Harvard Business School Press.

Castells M. (1996 [2000]) *Rise of the Network Society: The Information Age: Economy, Society and Culture*. (1996) 1st edition; (2000) 2nd edition. Blackwell Publishers Inc.

Castells M. and Hall P. (1994) *Technopoles of the world: the making of 21st century industrial complexes*. London, Routledge.

Christaller W. (1933 [1966]) *Central Places in Southern Germany*. (1933) 1st edition; (1966) translated edition. Englewood Cliffs, NJ, USA, Prentice-Hall.
Coase R.H. (1937) The Nature of the Firm. *Economica*, 4, 386–405.
Goddard J.B. (1973) Office linkages and location: A study of communications and spatial patterns in Central London. *Progress in Planning*, 1, 109–232.
Goddard J.B. (1975) *Office Location in Urban and Regional Development*. Theory and practice in geography. Oxford, Oxford University Press.
Goddard J.B. and Morris D. (1976) The communications factor in office decentralization. *Progress in Planning*, 6, 1–80.
Haig R.M. (1926a) Toward an Understanding of the Metropolis (Part 1). *The Quarterly Journal of Economics*, 40, 179–208.
Haig R.M. (1926b) Toward an Understanding of the Metropolis (Part 2). *The Quarterly Journal of Economics*, 40, 402–434.
Hall P. (1962) *The Industries of London since 1861*. London, Hutchison & Co.
Hall P. (1987) *Western sunrise: the genesis and growth of Britain's major high tech corridor*. London, Allen & Unwin.
Hall P. (1991) Moving Information: A Tale of Four Technologies. In Brotchie J., Batty M., Hall P. and Newton P. (eds.), *Cities of the 21st century: new technologies and spatial systems*. Chichester, UK, John Wiley & Sons.
Hall P. (1997) Modelling the post-industrial city. *Futures*, 29, 311–322.
Hall P. (1998) *Cities in Civilization*. New York, Pantheon Books.
Hall P. (2000) Creative cities and economic development. *Urban Studies,* 37, 639–649.
Hall P. (2003) The end of the city? "The report of my death was an exaggeration". *City*, 7, 141–152.
Hall P. (2005) Apologia Pro Vita Sua. URL http://www.balzan.org/en/prizewinners/sir-peter-hall/apologia-pro-vita-sua.
Hall P. (2014) Apologia Pro Vita Sua. in Tewdwr-Jones M., Phelps N. and Freestone R. (eds.), *The Planning Imagination: Peter Hall and the Study of Urban and Regional Planning*. Abingdon, UK, Taylor & Francis.
Hall P. and Markusen A. (1985) *Silicon landscapes*. London, Allen & Unwin.
Hall P. and Pain K. (2006) *The polycentric metropolis: learning from mega-city regions in Europe*. Earthscan.
Hall P. and Preston P. (1988) *Carrier Wave: New Information Technology and the Geography of Innovation, 1846–2003*. London, Routledge, Chapman & Hall.
Hall P., Markusen A., Osborn R. and Wachsman B. (1983) The American computer software industry: economic development prospects. *Built Environment,* 29–39.
Hutton T.A. (2004) The New Economy of the Inner City. *Cities*, 21, 89–108.
Hutton T.A. (2006) Spatiality, built form, and creative industry development in the inner city. *Environment and Planning A*, 38, 1819–1842.
Hutton T.A. (2008) *The new economy of the inner city: Restructuring, regeneration, and dislocation in the 21st century metropolis*. London, Routledge.
Lösch A. (1954 [1973]) *The economics of location*. (1954) 1st edition; (1973) 6th edition. New Haven, Connecticut, USA, Yale University Press.
Malecki E. (2014) The Information Age and Technological Change. in Tewdwr-Jones M., Phelps N. and Freestone R. (eds.), *The Planning Imagination: Peter Hall and the Study of Urban and Regional Planning*. Taylor & Francis.
Markusen A., Hall P. and Glasmeier A. (1986) *High Tech America: The what, how, where, and why of the sunrise industries*. Boston, USA, Allen & Unwin.
Markusen A., Hall P., Campbell S. and Deitrick S. (1991) *The Rise of the Gunbelt: the Military Remapping of Industrial America*. Oxford University Press.
Ozkul B. (2014) Changing home-to-work travel in England and Wales. *Regional Studies, Regional Science*, 1, 32–39.
Ozkul B. (2015) Von Thünen Revisited. *Built Environment*, 41, 99–111.
Polanyi M. (1967) *The Tacit Dimension*. Garden City, NY, USA, Doubleday Anchor Books.

Reades J. and Smith D. (2014) Mapping the 'Space of Flows': the geography of global business telecommunications and employment specialisation in the London Mega-City Region. *Regional Studies*, 48, 105–126.

Tanenbaum A.S. (1996) *Computer networks*. 4th edition. New York, Prentice Hall.

Taylor P., Hoyler M. and Verbruggen R. (2010) External Urban Relational Process: Introducing Central Flow Theory to Complement Central Place Theory. *Urban Studies*, 47, 2803–2818.

Tewdwr-Jones M., Phelps N. and Freestone R. (eds.) (2014) *The Planning Imagination: Peter Hall and the Study of Urban and Regional Planning*. Abingdon, UK, Taylor & Francis.

Vance J.E. (1970) *The merchant's world: the geography of wholesaling*. Foundations of Economic Geography. Englewood Cliffs, NJ, USA, Prentice-Hall.

Von Thünen J. (1826 [1966]) *The Isolated State*. Pergamon Press. With an Introduction by Peter Hall.

Weber A. (1909 [1969]) *Theory of the Location of Industries*. Materials for the Study of Business. University of Chicago Press, (1909) 1st edition; (1969) 7th edition.

Chapter 4
Transport and Place-Making: A Long View

Chia-Lin Chen

4.1 Introduction

Professor Sir Peter Hall was one of the most influential planners in modern Britain and highly respected worldwide. With his foresight and long-term vision—a holistic approach towards urban planning and development—transport was always an inseparably structural part of spatial planning strategies in his analysis of cities and regions. He often alluded to great transport heroes who transformed the world in different epochs. For example, the great Victorian engineer, Isambard Kingdom Brunel, the creator of the new London Underground, Frank Pick (with Albert Stanley and Lord Ashfield), and the father of the British motorway system, Sir James Drake. Furthermore, Peter Hall regularly referred to the renowned British economist Colin Clark's celebrated aphorism (1958) "Transport: Maker and Breaker of Cities" that traced the role of transport from early urban civilisation to the end of the 1950s to highlight the importance of transport for territorial development. Clark argued that the growth of cities have been shaped by the development of transport facilities that are dependent on the evolution of transport technologies. The relationship between transport technologies and cities is more complex—the transport system shapes the growth of the city while the previous development in cities constrains the availability of transport alternatives. Taking a long-term perspective, Peter Hall had been constantly revisiting this Clarkian path with new meanings to capture the symbiotic relationship between transport and cities and to seek innovative solutions for urban problems encountered from the 1960s up to the 21st century.

Throughout his prolific academic career spanning 55 years, Peter Hall was not only a forward-looking thinker and academic but also a knowledgeable historian of

Chia-Lin Chen, Dr., Xi'an Jiaotong Liverpool University, Suzhou, China, Lecturer in Urban Planning and Design; chia-lin.chen@xjtlu.edu.cn

R.D. Knowles and C. Rozenblat (eds.), *Sir Peter Hall: Pioneer in Regional Planning, Transport and Urban Geography*, SpringerBriefs on Pioneers in Science and Practice 52, DOI 10.1007/978-3-319-28056-1_4

planning ideas for urban development. He actively engaged in national debates and contributed to public policy and strategic plans. His influence has been wide-ranging and profound both domestically and globally. It is not the intention of this chapter to embrace all of Peter Hall's contributions to transport research. Instead, it focuses on the evolution of his thinking regarding the relationship between transport and the city—how the wider contexts, unexpected big events, and his personal experience had closely interacted to shape his thinking and contributions.

4.2 A Unique Generation: The Making of Modern London and Rebuilding Britain

Professor Hall was born in March 1932 and lived through a remarkable generation that shaped his vision for the future in three ways. Firstly, during his childhood, Peter Hall witnessed the "The Making of Modern London" (1919–1939) captured by Weightman/Humphries (1983). Tube stations at that time were symbols of modernity. These "modern palaces" were characterised by "huge, brilliantly lit entrance halls, the extraordinary indirect lighting on the escalators, the clean lines of the station platforms, the rivers of people flowing through the system" (Hall 2015: 1).

Peter Hall explained this childhood fascination with the London Underground in his inaugural lecture at UCL in 1994.

> …I must have been about two and a half when my father lifted me up to peep over the wall […] to see the Piccadilly line trains coming out of the tunnel […] I became instantly fascinated by the tube, and in particular by H. C. Beck's legendary map, which had made its first public appearance that very year, 1934. […] at the age of about five, I spent many obsessed hours with coloured crayons, copying it out. I suppose that in 1938, I must have been the only six-year-old in London who knew all the tube stations by heart (ibid.).

Secondly, Peter Hall's family relocated to Blackpool in 1938, the celebrated Lancashire seaside resort town, and during his post-war exile there, he received an "Etonian-quality" education. There always existed a revolutionary spirit deep in his mind captured well by this Marxist maxim: "the point was not merely to understand the world but to change it" (Hall 2014a: 270). Furthermore, in his adolescence he was immersed in "the extraordinary spirit, among everyone, to fight the war and then to rebuild a better world" (Hall 2015: 3). This special association with Blackpool during this extraordinary period might explain well why he was devoted to finding a strategic solution for Blackpool's revival in his last EU project—SINTROPHER (discussed later in this chapter).

Thirdly, while at the Geography Department at the University of Cambridge, Peter Hall was taught to believe that "geography could be actively applied to change the world" (Hall 2015: 6). Under the circumstance of these three main forces, one can begin to understand why the importance of strategic planning that underlies his vision and ambition was natural for him.

It can be said that transport had not been singled out as a professional subject until the early 1960s when traffic congestion was perceived as a problem to be solved within urban policy. In the wake of unprecedented increases in car ownership, mathematical models were drawn on to deal with the laws of traffic generation, distribution, modal split, and assignment, which regards traffic as a function of land use (Mitchell/Rapkin 1954). Underlying this modelling assumption is that with an analysis of socio-economic characteristics and activity patterns, future traffic demand could be estimated without doubt. Instead of producing estimated figures, what interested Peter Hall, or rather puzzled him, was the mutual relationship between transport and place-making—specifically, how the patterns of urban development and urban structure have been shaped by the transport networks and how this influenced human activities. Professor David Banister, a well-known expert in transport and planning and a very good friend of Hall's noted that Peter was not directly involved in the development of transport planning models or the transport planning process (Banister 2014). In Hall's terms,

> During my life, I have spent a lot of time thinking about how cities are structured through their networks, through their transport, and through ways in which innovation, enterprise and segregation are all influenced by such connectivities (Hall 2015: 20).

Set against this background, this chapter[1] focuses on the evolution of Hall's thinking and his profound contributions to the enhancement of the link between transport and urban development. Four major stages of his trajectory can be identified, namely his early contributions in the 1960s/1970s, the transformative period in the 1980s, revisiting the Clarkian paradox in the 1990s, and a synthesised period of strategic planning for polycentricity and addressing spatial inequality in the 2000s up to his passing.

4.3 Early Contributions

The first British motorway, the Preston bypass, opened in 1958, soon followed by a network of motorways in the 1960s. The consequent rise of car ownership had dreadful effects on urban traffic. Colin Buchanan's "Traffic in Towns" (1963) vividly reflected the public awareness of urban traffic congestion problems that were not yet foreseen by Clark (1958). Instead, Clark envisaged the impacts of private car transport on uniform urban sprawl and the decentralisation of homes and jobs, which led to him suggesting manageable sizes of community settlements in order to avoid "an ugly and planless dispersal of population" (Clark 1958: 250).

With the aim of tracing the influence of Clark's celebrated paper on Peter Hall's thinking, within the available literature, it can be seen that as early as in the mid-1960s, in the first edition of "The World Cities" (Hall 1966), Clark's "Transport: Maker and Breaker of Cities" (1958) and Gilmore's "Transportation and the Growth of Cities" (1953) were both cited in the final chapter "The Future of Metropolis".

---

[1]For Hall's contribution to transport and planning, Banister (2014) gives a concise review and summary.

Saying this, however they were not elaborated on in the main texts. The 1960s was a decade when large-scale/comprehensive planning was seen in terms of the task of renewing London's urban fabric for the motorway age. As Peter Hall described,

> all the long-term strategic planning of activities and traffic, has just become a major issue, perhaps the major issue, of the day; the newspapers and television screens were full of it (Hall 1969: 28).

Thanks to the encouragement of John Vaizey who "turned [Hall] from geography into planning", Hall's first planning book "London 2000"—a planning manifesto was published in 1963, imagining nearly 40 years ahead about what the future held for London, in terms of planning housing, jobs, transport and governance for growth.

Peter Hall regarded transport as a derived demand, arguing that "transport is not a primary problem: it only arises from the patterns of living, working and playing within the region. To reach an adequate solution, we need to go beyond transport, to the causes that bring it about" (ibid: 129). He investigated the metropolitan explosion by empirically analysing the growth patterns of the London metropolitan area, looking at jobs, population, homes, and travel. These formed the basis for building a New London and renewing the urban fabric. The key issue of transportation was said to solve the severe congestion caused by the increase of motor vehicles in large metropolitan areas. For this, he argued, "our urban reconstruction must be based on this fact" (ibid: 206).

Hence, one can argue that rising car ownership as well as building roads were the orthodoxy of the time. Hall's thinking about transport in the 1960s very much followed the trend, focusing on problem-solving rather than abstract ideals. "The solution we should be applying in these areas depends on the amount of traffic generated by land use" (ibid: 208). Furthermore, he highlighted the necessity of 'segregation' in order to accommodate cars.

> You need to keep vehicles away from where people eat and sleep and work and think and talk and shop; and put them where they belong, in pipes sealed off from the rest of the system, like bottled genies ready to perform the miracles we ask of them. At present, in an Aladdinesque nightmare, they have run amok (ibid: 206).

In the second edition of "London 2000" published in 1969, Hall made a balanced verdict on Buchanan's influential report, arguing that it was "over-praised and over-attacked" (ibid: 217). He recognised the critical value of the Buchanan Report lay in its assertion that the capacity of a city to take traffic should not be the traffic flow capacity but the environmental capacity—noise, smell, accidents and visual intrusion—whereas the report fell short in three aspects e.g. failing to produce a comprehensive guide to address the traffic problems, a confusing distinction between 'essential' and 'optional' traffic, and a very technical viewpoint that ignored other policy measures such as decentralisation. Hence, as ever, he was a fluent communicator and stressed two methods of segregation, a P (precinctual)-solution for where the traffic generation was light and a V (vertical)-solution for other occasions where pedestrians could navigate without distraction above while road traffic passed below. Meanwhile, Hall warned that a V-solution needed to be applied with great caution:

> The vertical solution will never be suitable everywhere; it is merely an element in an eclectic solution for London. The art of central planning will be the art of applying it sensitively; which will prove one of the major responsibilities of the next generation of planners (ibid: 151).

His first book chapter on transport and urban forms, entitled ‘Transportation’ (Hall 1970) was published in 1970 for an edited book, “Developing Patterns of Urbanisation”, where his analyses critically dismantled the complexity of urban forms by adopting a fundamental distinction between intra-urban and inter-urban transportation. At the time, although he was aware of the limitations of car transport for inter-city journeys, he justified using a single mode such as car for intra-urban transport. The central issue for urban and transport planners was the trade-off for future development up to 2000. He examined the possibility of allowing private transport in the growing outer peripheries of the greater metropolitan regions.

> …As city regions disperse on the regional scale, are they to reconcentrate on a local scale? How far should localised high density nodes of activity, linked to high density public transport but with inevitable congestion limiting private transport, be permitted and even encouraged in the growing outer peripheries of our great metropolitan regions? This, above all, is the central question for urban organisation in Britain, and in Europe, for the years up to 2000 (ibid: 156).

However, the increase of car ownership was not a manageable, constant rise; rather, it was taking place rapidly and in unprecedented terms. The reconstruction of major urban centres with new urban motorway networks did not effectively alleviate large-scale congestion; instead, the situation worsened progressively. Around the end of the 1960s and the beginning of the 1970s, in line with social and environmental concerns, a series of protests and campaigns organised by the political movement—“Homes before Roads”—brought about a subsequent policy change against building motorways and in favour of public transport and traffic management instead. These developments dramatically gave rise to the end of large-scale comprehensive planning through urban motorways. Hall recalled the moment when popular attitudes towards cars had fundamentally shifted.

> …From a belief in the whole-sale construction of cities around the car to a belief in conservation/preservation of existing cities limiting the impact of cars on cities to the maximum extent possible. It was one of the biggest and the most sudden psychological changes I have probably observed that ever occurred in the history of the 20th century.[2]

[2] *Source* BBC Documentary: “The Secret Life of The Motorway” Part 3, 14 min:28 s–15 min:05 s.

4.4 A Transformative Period: The Great Rail Revival and Return to the City

The late 1970s and the early 1980s—the end of the post-war era "Les Trente Glorieuses"—was a critical watershed in planning attitudes in many West European countries, moving from comprehensive planning towards small-scale, more sensitive approaches to planning. A few major planning decisions such as UK's third airport and the freeway package were controversially suspended. These big debates were analysed at the time and regarded as "Great Planning Disasters" (Hall 1980a, 1982). In it, Peter Hall defined two types of disasters, namely positive and negative disasters. Positive disasters are those "big, discrete, single-shot" projects that were implemented but were then felt to have been wrong, whereas "negative disasters" are those plans that were abandoned and not yet realised (Hall 1982: xix).

The aforementioned wider context and different circumstances played a key role in shaping Peter Hall's planning ideas. In response to the two controversial projects "London Ring Road" and "London Third Airport" and to make changes possible instead of endless debates, the approaches he adopted were incremental and avoided bulldozing, based on practicality and necessity. As he argued,

> …The right way to have built ring ways in London […] might have been not through a 347-mile, £200,000,000-plus package, but through gradually upgrading the old road and building stretches of new road […]. Similarly, […] for airport capacity […] [upgrading] an existing airport to the limits of its capacity […] That idea […] was very much a reaction to the euphoric, expansionist planning style that characterized the 1960s, and to the big projects that resulted. The peril in the 1980s, to judge from the experience of the similarly constrained 1970s, is almost the opposite: it is of doing nothing, or almost nothing at all (Hall 1982: xx).

Peter Hall extensively reviewed the uncertainty and irrationality of decision-making processes and questioned the role of different actors (the community, the bureaucracy, and the politicians) who were involved. Taking into account the fact that the state was not able to make any decisions without encountering controversial situations, he stressed that the "do nothing solution is destined to be the mistake of the 1980s" and represented "the disease we should be fighting" (Hall 1982: xx). Evidently, Peter Hall reflected closely on these controversial events; In "London 2001", published 25 years after "London 2000" was first published, he looked back at these urban discontents and admitted the political complications of the time:

> …The fact is that to achieve real and lasting improvement in London's traffic environment, someone has to be sacrificial victim; there is no way around it. When it came to the crunch, the politicians' nerves failed them, again and again. Thus does local democracy lead to a state of stalemate, where all suffer because no one may ever lose … I must confess that, much older and, I think, wiser, I have no easy solution to this dilemma (Hall 1989: 160).

Due to the severe congestion caused by the unprecedented increase of car ownership, the period from the 1980s onwards was an era of public transport revival that addressed the problematic externalities of private car transport.

Meanwhile, gradually, in addition to the large World Capital Cities, Peter Hall directed his concerns toward non-capital, regional cities and their transport patterns. In the edited book—"The Future of the British Conurbations Policy and Prescriptions for Change" (Hall 1980b)—he contributed the chapter "Transport in the Conurbations," looking at seven British post-industrial conurbations (London, Birmingham, Manchester, Glasgow, Leeds, Liverpool and Newcastle). Given the policy shifts taking place, first from road to rail, then from rail to more modest piecemeal improvements in "public transport," Peter Hall pondered what this meant to the transport policy in varying conurbations and whether transport should be a focus of public policy. He argued that the role of transport for cities had moved away from urban efficiency to urban equity, referring specifically to the transport problem for inner-city residents who did not own cars noting that they might be disadvantaged with the loss of jobs in the inner city jobs and the new jobs created were further away. As is typical, Peter Hall suggested viable alternatives—"perhaps subsidising reverse-commuting, or whether jobs can be re-created in the inner city" (ibid: 169–170).

It was gradually evident that Peter Hall had located a key issue that required addressing; without intervention, through the creation of jobs and homes through urban regeneration policies, improvements in transport on its own could not solve the problem.

> …How far the battle to improve public transport can ever be won in the face of rising car ownership… For one point is certain: that as long as the dispersal of homes and jobs continues, so will conventional public transport fight a losing battle against the flexibility of the private car (ibid: 170).

Similarly, Hall went on to highlight the disconnection between transport authorities and local authorities regarding the role of transport investment. This is alluded to in his case study on London's Jubilee Underground line through the Docklands, which while presented with a negative rate of return, the Greater London Council claimed it was vital to increase job opportunities by extending local labour markets (ibid). In retrospect, as recent research (Jones 2015) has shown, one can confidently argue that the Jubilee Line extension and London Dockland Light Rail have successfully functioned as catalysts for the regeneration of the London Docklands beyond transport-related benefits. This case demonstrated that transport appraisal approaches which are narrowly focused on particular criteria failed to take wider benefits into account. However, this also raised another question about different characters of different places—particularly in the context of widened inequality between places such as London compared to other core cities in the UK.

4.4.1 "Can Rail Save the City?"

In the 1980s, Peter Hall's attitude towards transport also gradually shifted from car to public transport due to the influence of Professor Carmen Hass-Klau, who was initially one of Hall's master's students and then research assistant and one of his Ph.D. students at the University of Reading. As Hall acknowledged in the symposium on 25/26 June 2012 in UCL that, "…She (Carmen) did have some quite fundamental influences on my thinking, demonstrating to me the value of positive public transport investment in especially German and other continental cities that did make the shift evident in London 2001" (Chen 2015: 115). Indeed, when "London 2001" was published in 1989, urban transit systems had returned to be a key player as a few new rail schemes were being implemented (Thameslink opened in 1988 and Docklands Light Railway in 1987) in line with the rail plan produced by a committee under Sir David Barran in 1974 (Greater London and Department of the Environment 1974). In addition, unlike "London 2000" where US examples were mainly referred to, in "London 2001", Hall studied good examples from continental European countries and beyond. For instance, Hall recommended the Paris regional express rail system (RER) as an outstanding example for London (Hall 1989).

Professor Hass-Klau's personal experience with public transport in the childhood helped her develop a positive attitude towards it. In her words,

> …At this time, I already thought that I like cities without cars and I experienced in Germany large-scale pedestrianisation which started in Munich and was copied in many other German towns and I also had positive relationships to the public transport from my childhood. We didn't have a car. I had to use buses and trams to go to school. I love trams. That's my part of my adolescence and my adulthood. So I could not understand that somebody is so negative about public transport… (interview, 24 April 2015).

Professor Hass-Klau also recalls that it was with the book "Can Rail Save the City?" (Hall/Hass-Klau 1985) that Hall developed a change in attitude. However, it was a slow process geared towards the understanding of the value of public transport. When the book was finished, she had been convinced by the importance of public transport. Hall was still not convinced, which is revealed in the conclusions.

The background of this book was a major shift in transport planning policies for both Germany and the UK from planning for free car usage towards the promotion of public transport. At that time, German cities—relatively medium-sized—were building new transit systems (21) generously while British cities (3) had built relatively little. Meanwhile, almost all German cities pedestrianised central shopping streets with a whole network of streets in contrast to largely fragmented pedestrian shopping precincts created in British cities.

The combination of the two policies i.e. public transport and pedestrianisation did seem to work in concert to enable German cities to create car-less/high quality city centres. Coincidentally, German cities had more successfully retained cities' positions as employment centres than British ones. Therefore, Hall and Hass-Klau wanted to know whether transport policies had helped local economies. Interestingly, the findings showed that increases in passenger levels were not followed by increases in economic activity in Germany. Similarly, a significant impact in the UK was unable to be established. Saying this, it is true that it is very difficult to isolate the effects of transit from other causes. The importance of transport on economic impacts has only been recognised very recently through theories related to agglomeration economies. Evidently, his conclusion showed his reserved attitude towards the value of public transit investment.

> … the processes of urban growth and decline […] so far imperfectly understood. Transit investment is in large measure irrelevant to these processes, though it may affect some of them at the margin. Rail cannot save the city […] because the forces that are taking it down are far wider and far deeper than mere questions of accessibility. That is not to deny the potential importance of transport investment to the regeneration of a city's economy. It is to say that they would need to be planned in the context of a far better understanding of that city's malaise (ibid: 169).

To this conclusion, Professor Hass-Klau indicated that "We could not establish an increase in the economic activities because there was no small scale data available at that urban level. The only way we could have proved that would have been by interviewing retailers" (email exchange, 31 August 2015). In her new book "The Pedestrian and the City" (Hass-Klau 2015), the overwhelmingly positive impact of pedestrianization on retailing was evidently presented. She further stressed a holistic approach that "The issue was not only investment in public transport but also investment in urban design and the combination of both was (and still is) the key to success" (email exchange, 31 August 2015).

Peter Hall's personal experience also played a key role in changes in attitude. As Carmen Hass-Klau suggested that the early 1980s was a particular point in Peter Hall's life when he worked in UCL, and first starting using a Travel Card. Hall thought this was brilliant and became a totally convinced public transport user. This is in sharp contrast to the life he experienced in the US where he drove up to 10 h a day in 1987. While in London, using cars is simply not a good idea for people working in central London unless one lives outside and wants to go from outside to outside (interview, 24 April 2015).

4.5 Revisiting the Clarkian Paradox

In the early 1980s, Peter Hall joined UC Berkeley, a move that brought him extraordinary experience. With close proximity to the power house of information technology in Silicon Valley, he was intellectually inspired and his planning ideas were enriched as he was fascinated by the role that technology played in urban planning theories, going on to collaborate closely with a group of international researchers for nearly 15 years on a number of projects related to technology and the city (Brotchie et al. 1985, 1987, 1995, 1999; Brotchie 1991).

The notion of long waves in history—economic fluctuations every 50–60 years—explains cyclical patterns of capitalism through external technological change and subsequent commercialised innovations. He was particularly drawn by Schumpeter's theory of innovations that advanced Kondratieff's long wave theory by making a distinction between the role of innovations and inventions in the propulsion of economic restructuring (Schumpeter 1964, 1982 [1939]). These innovations, the engine of "creative destructions" result from "new men", entrepreneurs' innovative commercialisation of technological inventions in all aspects of the economic realm constantly applying technological advances for new methods of production, new products, transport, newly created markets and new forms of industrial organization.

This period represented a highly productive and inspirational part of Peter Hall's academic trajectory. "Silicon Landscapes" (Hall/Markusen 1985), "Western Sunrise: The Genesis and Growth of Britain's Major High Tech Corridor London" (Hall et al. 1987), "High-Tech America" (Markusen et al. 1986), and "The Rise of the Gunbelt" (Markusen et al. 1991) were his key intellectual contributions during this period. How Hall's thinking regarding technology and cities contributed to the concept of smart cities is examined in more detail by Professor Mike Batty (2016) in this volume.

These inspirational experiences sparked a renewed appreciation and innovative understanding of Clark's aphorism. The combination of the long-wave theories and his revisiting of Clark's "Transport: Maker and Breaker of Cities" (1958) was an innovative idea. Peter Hall found resonance in Professor Brian Berry's work (Berry 1991) that attributed cycles of urban development to advances in transport or communication technologies. He exemplified his point with three key historical events, namely the first railways of the 1830s appearing at the beginning of Middlemarch; the surge of modern urban metro and underground systems around 1900—when the third Kondratieff just took off; the escalation of motorway construction in the late 1950s signifying the starting point of the fourth (Hall 2015).

Since then, Hall often quoted Clark's classic paradox to highlight the symbiotic but complex relationship between transport and the city—"[T]he two could get out of step, and indeed very often did so" (Hall 1994: s79). By conducting a historical review, he identified that cities in the developed countries have experienced four successive crises of transport technology and urban form with the fourth is still in progress.

A paper published in 1992 entitled "Transport: Maker and Breaker of Cities" (Hall 1992) was the first of his serial contributions to the topic (Hall 1994, 2014b). Over these years, Peter Hall's analyses of urban forms and solutions implemented in many cities had been refined with sensible responses to the shifting opportunities and the wider context within which he was situated. He traced how nearly 200 years of urban forms were shaped by transport technologies from pre-public transport date up to the present day. By 1850, pre-public transport cities that developed essentially on foot experienced their first crisis: cities could not grow further. Horse trams and steam-hauled commuter railways were developed to create streetcar suburbs that gave rise to "early public transport cities" around 1890. Soon after, a second crisis arose due to limitations on further growth in the large cities. Then a new phase around 1890–1900 began to shape "late public transport cities"—through electrified trams, commuter trains, underground electric railways, and then radial line extensions above ground to serve wider catchments of new suburban rings by the 1920s and 1930s. The third crisis took place in most cities in developed countries around the 1960s (in Los Angeles, it was as early as the 1920s) when the rise of cars conflicted with the existing urban fabric.

Peter Hall concluded that three kinds of cities were formed to react to this development. The first kind, and the most well-known, was Los Angeles' answer to the crisis, deliberately letting public transport decline thus resulting in an auto-oriented city. Such a model was also followed by other US cities. He drew on Thomson's (1977) classifications to stress another two kinds of urban forms in the North American and European cities, namely strong and weak centre cities. Strong centre cities were New York, Paris, London and Tokyo that were extremely rail dependent whereas weak centre cities referred to regional cities in US and Europe with limited commuter rail services. Strong and weak centre cities, as Peter Hall reminded us, were hybrids—places outside their dense cores were car-dependent and there always existed a mixture of other public transport modes, buses, trams, light rail, bus rapid transit (BRT) etc. As a result, existing public transport radial systems could not adequately cater for the cross-flows because the decentralisation of homes and jobs was followed by a shift to cross-commuting dominated by private cars (Hall et al. 1993). What is more, Peter Hall accentuated two contemporary contradictions, namely relating to congestion and energy consumption and argued these constitute a fourth crisis in the organisation of urban transport.

This fourth crisis was faced by many cities worldwide. He reviewed a wide range of studies (completed and in progress) with three major aspects of measures in practice to address nonconventional (suburb-to-suburb) and polycentric "spread city" patterns underlaid by car dependency, namely traffic restraint, road pricing, and combining transport and land use policies (Hall 1994). In addition, he highlighted that integrated transport and land use planning policies should be implemented on multiple spatial scales that are interlinked. At the local level, he raised examples of experiments associated with pedestrian-friendly communities that took place in the Californian cities of Sacramento and San Jose. At the metropolitan level, Peter Hall noticed the development of new orbital rail systems (RER) in Paris whereas elsewhere some were experimenting with Bus Rapid Transit systems or paratransit systems which proved more appropriate and effective in their contexts as good alternatives to cars. However, based on limited evidence, he stressed that 'it is not clear …. how to develop effective sustainable urban forms at larger sub-regional and regional scales' (Hall 1994: 92). At the inter-urban level, high-speed rail including the Maglev was mentioned as potentially revolutionary for inter-urban transportation over high volume traffic corridors.

Twenty years on, many more cities have been very much following these solutions. Crossrail in London, taking a French RER model, finally got approved and expects to be completed by 2018. The network of High Speed Rail (HSR) routes continues to grow worldwide. The development of the Tokyo-Osaka Maglev system has continued to be developed. In April 2015, a new speed record was set at 603 km/h. In 2045, the train time between Tokyo and Nagoya will be reduced to 1 h 7 min from the current 2 h 25 min (McCurry 2015). However, more research will be needed to understand how effective these solutions are in different contexts.

By the early 1990s, inspired by the long-wave theories, Peter Hall pondered that "cities change under the influence of their transport systems and that these change under the impact of technological change, it would be expected that some dramatic shifts in technology might occur" (Hall 1992: 274). Saying this, he consistently urged for new transport technological innovations and their application in cities. He expected that a 21st century metropolitan transport system would see the injection of information technology into the operation of various modes of transportation, alluding to the possibility of full or partial automation. Possible alternatives include "small van-like vehicles, rather like the airport shuttles over either ordinary streets or an automated guide-way system" (ibid). Additionally, he also estimated that there will be a more collective or shared modes such as vans and car pools in the future. Many of these ideas indeed have been implemented in many cities and occasions and assuming prevalence.

4.6 The Synthesised Period: Strategic Planning for Polycentricity and Addressing Spatial Rebalancing

From the 1990s until his death, Peter Hall was able to synthesise his knowledge, wisdom and experience gained over his previous 30 years, With the advanced development of informational communication technology and the improvement in transport connectivity, processes of decentralisation and re-concentration operate within increasingly expanding city-regions to place limits on the process of decentralisation, especially regarding critical time limits to and from the central city. The POLYNET study (Hall/Pain 2006) suggested that improved accessibility of both 'e-' and physical links between major centres and secondary centres could crucially assist a monocentric region to develop into a better-integrated Mega-city Region (MCR).

Therefore, Peter Hall demonstrated to us how planning a poly-centric urban form that addresses spatial rebalance was a key problem to be addressed for cities and regions at both inter- and intra-regional levels in the 21st century and how the opportunities for transport technologies should be seized to assist this process.

In compliance with this thinking, he was closely involved in planning and designing inter-city high-speed rail (HSR) links both in the UK and the US, and intra-regional rail transport links within mega city-regions such as London Overground and for smaller peripheral sub-regions such as tram-trains. The latter, though financially expensive, was an innovative solution through the improvement of the South Fylde Line to regenerate Blackpool which had been suffering from structural changes and poor transport connectivity.

4.6.1 *High-Speed Rail as One of Two Driving Forces Shaping Urban Form*

Peter Hall had been looking for new transport technologies that could drive urban change. Although there had been a general lack of major technological progress, his answer to this was that "there will be high-speed trains, information technology" were quite evident through his belief and passion for HSR development. He described his intuitive feeling of an electric moment that reminded him of Stockholm in 1955 or Los Angeles in 1966 or Hong Kong in 1975 while riding on the TGV Atlantique from Paris to Le Mans

> …[W]hen the train emerges from a long tunnel and begins to run parallel to the motorway. Suddenly, you see a line of broken-down trucks. But they aren't broken down: they are travelling at 60 mph. And then you pass the cars, travelling at 80 or 85, at two and a half times their speed, leaving them somehow suspended weirdly on the highway ….The point is this: when you actually experience this, you absolutely know […] here is something new and definitively different, that is going to change our world beyond any possibility of doubt.

From this experience, he was totally convinced that this kind of key events that will lead to a structural change should be well appreciated and planned for the future (Hall 2015).

Not only an inspirational academic, Peter Hall was also a strategic, forward-looking planner and campaigner. In California, instead of purely debating the possibility of creating a HSR system, he led a group of students working on the layout of HSR plans. As Teitz asserted "it was an example of Hall's ability, not only to see, but also to do what others only imagined" (Teitz 2014: 439). In addition, Hall organised a research team in Berkeley working on reviewing the developmental impacts of HSR through empirical case studies generating lessons for California in the process (Sands 1993). It was in 2014 after thirty years had passed that the California HSR system was approved and the construction of the first section had begun. Teitz described a particular encounter:

> One day in the 1980s, as I was going to my office in Wurster Hall, which required walking through a busy studio, I noticed that the drafting tables were covered with large maps of Central California, overlaid with lines in various directions. Intrigued, I asked the students what they were doing. They answered that they were designing a high-speed rail (HSR) system for California.… Characteristically, Hall saw that it needed substance, so he and his students designed the layout of the system virtually as it would appear in the subsequent decades (Teitz 2014: 439)

After Peter Hall returned to the UK and began his professorship at UCL, he was critically involved in a series of strategic planning initiatives responding to HSR opportunities for urban and regional development. In the early 1990s, as a policy advisor to Secretaries of State for the Environment (1993–1994), Hall developed the strategy for the Thames Gateway sub-region, with the proposed Channel Tunnel Rail Link (later High Speed One or HS1) as its centrepiece. In addition, potential sub-growth poles—Stratford and Ebbsfleet—were suggested to trigger economic regeneration in the Thames Gateway region and to offer attractive connections to Central London, Southeast England and Northern Europe. The effects were significant. Owing to the concerns for sub-regional regeneration, the existence of HS1 and Stratford station, then under construction, was a principal consideration in the International Olympic Committee's choice of London, announced in 2005, as the location for the 2012 Olympic Games (Faith 2007). During 1994–1995, he advised the London and Continental Railways Consortium (LCRC) on planning and regeneration aspects of the Channel Tunnel HSR Link, however, the process was not straightforward or free of controversy and straightforward. Eventually, "The Right Line" (ibid) was selected thus LCRC was able to build and operate it. As he described himself, with these projects on track being realised, this period in some ways was the most satisfying of all (Hall 2015).

Peter Hall regarded the HSR investment "far more than just a transport engineering question critically, it's a strategic planning question" (Hall 2005: 46). In the paper reviewing the demonstrated effects of HSR in Europe and other countries (Hall 1995), he highlighted the developmental effects—a substantial CBD has been established on a green field over a period of 30 years since the arrival of HSR in 1964. Having said that, he was acutely aware of the uncertainty. He raised examples of Heathrow airport in 1943 and London Docklands in the 1980s to illustrate the difficulty of predicting the long-term development potential—in contrast to a modest appraisal the change of land use change and value lift is apparent. He argued that it was not surprising to see the estimates of the developmental potential of HSR stations had differed profoundly. He maintained that involving in a new technology and public-private partnership, "any forecasting exercise is likely to approach the limits of the possible" (Hall 1995: 85), which reflected his belief in interventions and strategies of HSR despite its uncertain nature.

Before HS1 was completed, the idea for future High-Speed Two (HS2) was in the making, with Peter Hall actively engaged to seize the next HSR opportunity. In the key article (Hall 2005), he developed the argument that planning strategies should be based on the distinction between "Regional Metro" high-speed services and "Very High-Speed" services. Given that the growth of rail use that has taken place since 1994, he proposed that bypassing these development corridors to connect directly to regional core cities in the Midlands and North would free space on the existing lines to accommodate traffic growth without congestion.

In parallel, findings from Hall's EU Polynet project (Hall/Pain 2006) provided vital theoretical and practical evidence to making the case for HS2. Later, he sketched an outline for a future HS2 network using the old trackbed of the Great Central Railway (Hall 2005). In addition to this he joined the Advisory Group to Greengauge 21, a non-profit advocacy group that published its Manifesto for a future HS2 network in January 2006, attracting widespread media coverage and stimulating major national debate.

Following these active campaigns, a series of follow-up government reactions and national transport policies took off. The HS2 Ltd was established by the Department for Transport in 2009 to examine the case and present a potential route, and in March 2010, the Labour government published a HS2 White Paper.

Similarly, in January 2012, the then incoming Coalition Government announced the decision to proceed with the construction of the line. There are two phases to the development, namely Phase 1 from London to the West Midlands in operation by 2026 with Phase 2 London to Manchester (and other northern cities) by 2032.

Although HS2 investment received cross-party support, it is very controversial. Given the existence of polarisation among supporters and opponents for mega infrastructure planning, similar to the time of the "Great Planning Disasters," Peter Hall closely examined the great debate and major points of contention of the HS2 proposal (Hall 2013). In response to the worsening issues related to widening inter- and intra-regional inequality, Hall's research and analyses continued to provide important evidence in response to issues related to "rebalance north-south divide" and Pierre Veltz's notion of the archipelago economy (Veltz 1996), through his recent research with the present author. They focused their analysis on the 30-year experience of InterCity 125/225 for inter-regional impacts (Chen/Hall 2011) and the positive economic impacts of HSR on the Lille *Métropole* sub-region (Chen/Hall 2012). In addition, Hall actively collaborated with local experts to propose High Speed North, a development that would connects cities in post-industrial cities of Northern England (Hall et al. 2014). This proposal was later adopted by the central government, lending their support to High-Speed Three for linking together the Northern powerhouse. This concept of a "Northern Powerhouse" with a dedicated minister appointed was announced by the Coalition British Government to boost the northern economy for addressing the widening regional inequality "north-south divide". Strengthening the transport links across the northern core-cities from Liverpool to Hull (including Manchester, Leeds, Sheffield, and Newcastle) is expected to enable a more balanced economy—the northern regions vs. London and the South East. This is especially critical since because Transpennine rail services between these northern cities have been inadequate and very slow and there is a critical need for better services in terms of reliability, speed, and direct service patterns. However, this commitment from the government is subject to the financial uncertainty. The electrification of the key rail route Manchester-Leeds was paused in summer 2015 (Pidd 2015).

4.6.2 Planning a Poly-centric Urban Form for London and South East England

Apart from HSR that would serve inter-urban travel, in order to address the issue of cross-commuting outside the radial rail network, Peter Hall contributed to the concept of growth corridors in extended mega-city regions (Hall 1989, 1995) and developed the "Regional Metro" concept. Further, in a report for the RTPI in 1999, in order to address the issue of unbalanced development, against a backdrop of the growth of advanced service employment in the central area (Zone 1), and stagnation in middle and outer ring sub-centres, leading to increased strain on radial rail

Fig. 4.1 *Note* New 5-car trains introduced in November 2014 and an Overground train was named after Professor Sir Hall to recognise his valuable work and contribution on 30 April 2015 held by Transport for London (*photo credit* Chia-Lin Chen)

services into the centre, Hall developed the concept of a polycentric London, based on completion of an Orbirail system (now London Overground) through the capital's middle ring (Hall et al. 1999). This plan was not straightforward. Initially, it was embodied in the first drafts of the Mayor's 2004 London Plan when Hall was the chairman in a Mayoral Advisory Committee. Once it was subsequently abandoned by the then Mayor and was revised by Boris Johnson after his 2008 election; Orbirail fully opened in December 2012, London Overground proved to be a big success. In 2012, London Overground carried 120 million passengers—nearly four times the number carried when it launched in 2007. Including the Orbirail extensions, demand has increased 280 %, quadrupling from 2.57 to 9.83 m journeys per four-weekly period. A £320 m programme of introducing 5 car trains to increase capacity by 25 % was implemented to satisfy the growing demand (Fig. 4.1).

4.6.3 Improving Transport Connectivity for Peripheral Sub-regions and Regeneration

In his later years, uneven urban and regional development widened, generating serious socio-economic problems, while issues surrounding uneven infrastructure investment and other financial challenges loomed large. Meanwhile, a wide range of transport technologies (conventional rail/high-speed rail, electronic suburban tram, car, metro, light rail) have progressively developed up to the point that a mega polycentric urban structure had evidently emerged relating to both forces of spatial decentralisation and centralisation. This is especially the case in the Mega-city region developing around London but a similar pattern can also be applied to other cities worldwide. Therefore, beyond a single dominant mode, the issue lies in how different transport modes can exploit their own strength and collectively shape dynamic urban territories engendering more just and sustainable development.

The POLYNET project has demonstrated that urban structures in the UK and Europe have become increasingly polycentric, both physically (morphologically) and functionally, with individual towns and cities linked by fast, high-quality public transport which encourage agglomeration, networking and clustering. However, in practice this process has limits: too easily, high-speed networks can encourage the growth of larger cities and their regions at the expense of other urban locations, particularly those experiencing painful adjustment from a manufacturing (goods-handling) economy to a knowledge based (cognitive/cultural) economy. Peter Hall typically responded to this development, advocating the case for improved transport connectivity both between and—increasingly—within regions, leading to regional and local transport investment in North West Europe in two major ways.

First, further work in the UK by Green/Hall (2009) systematically examined the provision of station services in a wide range of rail interchanges and accordingly produced a priority list for station hub investment for the Secretary of State for Transport, which led to an investment programme most notably of £44 million at Manchester Victoria station (see Fig. 4.2).

Secondly, the EU INTERREG IVB project—SINTROPHER[3]—consisting of five major partner countries and costing €23 million addresses the issues of transport connectivities of peripheral regions by investing in intra-regional networks which link peripheral regions with their (sub-)regional centre so as to spread the benefits of accessibility from successful core city hubs into surrounding regions (Hall 2014b: 282). Although the final research output of this 6-year project is yet to be published (expected in September 2015), these studies have proved extraordinarily instructive in understanding how transport investments can be co-ordinated to produce spatial benefits—and, equally important, how difficult it can be to extend best practice from a few pioneering European places to the other places whose economies may be struggling due to their structural problems in addition to their poor transport connecivities. For instance, a parallel comparative case study of two regions—Manchester and its sub-regions in North West England (UK) and Lille and its sub-regions in Nord-Pas-de-Calais (France)—showed that, although in both cases faster train connections with the national capital did provide economic benefit

[3]SINTROPHER (Sustainable Integrated Tram-Based Transport Options for Peripheral European Regions) and SYNAPTIC (Synergy of New Advanced Public Transport Solutions Improving Connectivity in North–West Europe).

Fig. 4.2 The newly renovated Manchester Victoria Station in July 2015 (*Photo credits* Chia-Lin Chen)

to two regional capitals, the same was not true for some other sub-regions around, especially those post-industrial sub-regions (Chen/Hall 2012). That said, it demonstrated the need for positive public intervention in new transport links to 'irrigate'—that wonderful French word—those languishing regions (Hall 2014b: 282). Concerning responses to new transport opportunities, the contrast between the British and French cases in multi-level public interventions was manifest through a qualitative investigation—closely examining the developmental processes at different levels that critically addressed spatial inequalities (Chen/Hall 2013, 2015 and Hall/Chen 2013).

4.7 Conclusions

Sixty years have passed, but what Clark had predicted about the anti-urban impact of motorisation in the end of the 1950s,—"the complete disintegration of the city, even of the conurbation" (Clark 1958: 250) did not happen in reality. Clark did not anticipate location theories of agglomeration economies and the prevailing division of labour that are characterized in the post-industrial society. In the era of knowledge economy, instead of uniform urban sprawl, a polycentric urban form has been reinforced through forces of decentralisation and centralization. Understanding this complex relationship was characterised and captured well in Peter Hall's work.

Under the banner of Clark's "Transport: Maker and Breaker of Cities", Professor Hall had closely analysed patterns of urban change from the 1960s onwards up until his passing, greatly contributing to a better understanding of the relationship between transport and the city. Tracking his evolution, his thinking on the role of transport evolved from the initial focus on urban efficiency and urban equity at the metropolitan level, to a planning-led approach focused on spatial strategies at both inter-regional and intra-regional level. At the same time, his research focus shifted from mainly large metropolitan areas to much wider regional territories (both conurbations and peripheral areas). Throughout his life, Hall persuasively demonstrated that new opportunities need to be seized, proactive actions needed to be taken, and new meanings need to be invested according to the societal change, economic restructuring, and technological breakthrough.

For Peter Hall, the relationship between transport and development is interwoven and inseparable in creating an overall vision of good cities for better lives. In practice, integrating transport and urban planning has been difficult to achieve. London, the capital of the UK, a large metropolitan area, was the first city that Peter Hall's major academic contributions focused on. His enthusiasm and self-driven sense of mission led him to study many more places domestically and internationally. He constantly looked for new inspiration and better practice that offered potential solutions to urban issues emerging in his deep-loved British soil. By comparison, his research on good practice from different parts of the world provided critical insights into the complexity of urban development, enabling a deeper understanding of the real world wherein a variety of underlying factors interact to shape different paths of development i.e. political climate, leadership, technological changes, regulations, value systems, historical events, economic trajectories etc.

He reminded policy-makers of the continuous challenges and widened inequality many cities were facing. A better understanding of this, he believed, enables real action and progressive improvements. As he wrote in his recent article Apologia pro Vita Sua,

"Good cities for better lives are not a distant or impossible dream. But we need to find out how, from those places and those people who have learned how and can show us how. That's what real research, in and on real cities, is all about. To return to that long-ago saying of Marx, the point is indeed to understand the world—but then to use that understanding, however modestly, to improve it" (Hall 2014b: 282–283).

At present, facing unabated urban challenges as well as new opportunities, the mission is greater than ever. We, the young academics, should keep his spirit in mind and take actions. As Peter Hall accentuated, through better understanding and good intentions, we can make a better future world for our and future generations.

Acknowledgement The author would like to express sincere gratitude to Professor Carmen Hass-Klau, Professor Michael Hebbert and Dr Nick Falk who generously spoke to me about their understanding and observation of Peter Hall on the topic of this subject. Special thanks for valuable comments on my draft from Professor Hass-Klau, Professor Hebbert, and the editor of this book, Professor Knowles. Also, thank you to Peter Hall's prolificacy throughout 55 years of an excellent academic career. It is through his writing and my immense personal privilege working closely with him on my Ph.D. and later the EU INTERREG IV project that was led by him until his death in July 2014, that it is possible for me to understand better his ideas and passion on this subject and his passion for change.

References

Banister, D. (2014). Transport and Planning. *The Planning Imagination: Peter Hall and the Study of Urban and Regional Planning*. M. Tewdwr-Jones, N. A. Phelps and R. Freestone. London, Routledge: 204–214.

Berry, B. J. L. (1991). *Long-wave Rhythms in Economic Development and Political Behavior*. Baltimore, The Johns Hopkins University Press.

Brotchie, J. F. (1991). Fast Rail Networks and Socioeconomic Impacts. *Cities of the 21st Century: New Technologies and Spatial Systems*. J. F. Brotchie, M. Batty, P. Hall and J. Newton. New York, Longman Cheshire: 25–37.

Brotchie, J. F., Newton, P., Hall, P. and Nijkamp, P., Eds. (1985). *The Future of Urban Form: The Impact of New Technology*. London, Croom Helm.

Brotchie, J. F., Brotchie, J. F., Hall, P. and Newton, P., Eds. (1987). *The Spatial Impact of Technological Change*. London, Croom Helm.

Brotchie, J. F., Batty, M., Blakely, E., Hall, P. and Newton, P., Eds. (1995). *Cities in Competition: Productive and Sustainable Cities for the 21st Century*. Melbourne, Longman Australia.

Brotchie, J. F., Newton, P., Hall, P. and Dickey, J., Eds. (1999). *East West Perspective on 21st Century Urban Development: Sustainable Eastern and Western Cities in the New Millennium*. Aldershot, Ashgate.

Buchanan, C. and Crowther, G. (1963). *Traffic in towns : a study of the long term problems of traffic in urban areas : reports of the Steering Group and Working Group appointed by the Minister of Transport*. London, Ministry of Transport.

Chen, C.-L. (2015). "Peter Hall: Inspirer and Innovator of Railways and Regions." *Built Environment* 41(1): 112–128.

Chen, C.-L. and Hall, P. (2011). "The impacts of high-speed trains on British economic geography: a study of the UK's InterCity 125/225 and its effects." *Journal of Transport Geography* 19: 689–704.

Chen, C.-L. and Hall, P. (2012). "The wider spatial-economic impacts of high-speed trains: a comparative case study of Manchester and Lille sub-regions." *Journal of Transport Geography* 24: 89–110.

Chen, C.-L. and Hall, P. (2013). "Using High Speed Two to Irrigate the Regions." *Built Environment* 39(3): 355–368.

Chen, C.-L. and Hall, P. (2015). High-speed Trains and Spatio-Economic Impacts: A British-French Comparison on Two Scales: Intra- and Inter- regional. *International Handbook on Transport and Development* R. Hickman, M. Givoni, D. Bonilla and D. Banister. London: Edward Elgar.

Clark, C. (1958). "Transport: Maker and Breaker of cities." *The Town Planning Review* 28(4): 237–250.

Faith, N. (2007). *The Right Line: The Politics, Planning and Against-the-odds Gamble Behind Britain's First High-Speed Railway*. Kingston, Segrave Foulkes.

Gilmore, H. W. (1953). *Transportation and the growth of cities*. Glencoe, Ill., Free Press.

Green, C. and Hall, P. (2009). *Better Rail Stations*. An independent review presented to Lord Adonis, Secretary of State for Transport. London, Department for Transport.

Hall, P. (1966). *The World Cities* (1st edition). London, Weidenfeld & Nicolson.

Hall, P. (1969). *London 2000* (2nd ed.). London, Faber.

Hall, P. (1970). Transportation. *Developing Patterns of Urbanisation*. P. Cowan. Edinburgh, Oliver and Boyd: 130–157.

Hall, P. (1980a). *Great Planning Disasters*. London, Weidenfeld and Nicolson.

Hall, P. (1980b). Transportation. *The Future of the Conurbations*. G. Cameron. London, Longman: 146–171.

Hall, P. (1982). *Great Planning Disasters*. Berkeley, University of California Press.

Hall, P. (1989). *London 2001*. London, Unwin Hyman.

Hall, P. (1992). Transport: Maker and breaker of cities. *Environmental Issues in the 1990s*. A. M. Mannion and S. R. Bowlby. Chichester, Wiley: 265–276.

Hall, P. (1994). "Squaring the circle: can we resolve the Clarkian paradox?" *Environment and Planning B: Planning and Design* 21(Supplement): s79-s94.

Hall, P. (1995). A European perspective on the spatial links between land use, development and transport. *Transport and urban development*. D. Banister. Oxford, E&FN Spon: 65–88.

Hall, P. (2005). "A low-cost high-speed rail line for the north." *Town and Country Planning* February: 46–48.

Hall, P. (2013). "High-Speed Two-The Great Divide." *Built Environment* 39(3): 339–354.

Hall, P. (2014a). Apologia pro Vita Sua. *The Planning Imagination: Peter Hall and the Study of Urban and Regional Planning*. M. Tewdwr-Jones, N. A. Phelps and R. Freestone. London, Routledge: 269–284.

Hall, P. (2014b). *Good Cities, Better Lives: How Europe Discovered the Lost Art of Urbanism*. Abingdon, Routledge.

Hall, P. (2015). A History of Contemporary Planning: Personal Stories, Personal Journeys in Britain and America. *Encounters in Planning Thought: Autobiographical Essays from Key Thinkers in Planning*. B. Haselberger and L. Saija, (Forthcoming).

Hall, P., and Chen, C.-L. (2013). Using HS2 to irrigate the Regions. *Town and Country Planning*, 82, 170–174.

Hall, P. and Hass-Klau, C. (1985). *Can Rail Save the City? The Impacts of Rail Rapid Transit and Pededtrianisation on British and German cities*. London, Gower.

Hall, P. and Markusen A., Eds. (1985). *Silicon Landscapes*. Boston, Allen and Unwin.

Hall, P. and Pain, K. (2006). *The Polycentric Metropolis- Learning from Mega-city Regions in Europe*. London, Earthscan.

Hall, P., Breheny, M., McQuaid, R. and Hart, D. (1987). *Western Sunrise : The Genesis and Growth of Britain's Major High Tech Corridor*. London, Allen & Unwin.

Hall, P., Edwards, M., and Drummond, R. (1999). *London's spatial economy : the dynamics of* change : for London Development Partnership. London, London Development Partnership.

Hall, P., Sands, B., and Streeter, W. (1993). *Managing the suburban commute: a cross-national comparison of three metropolitan areas*. WP-596. Berkeley, CA, Institute of Urban and Regional Development, University of California at Berkeley.

Hall, P., Thrower, D. and Wray, I. (2014). "High-speed north- Building a trans-pennine mega-city." *Town and Country Planning* 83(4): 160–167.

Hass-Klau, C. (2015) *The Pedestrian and the City*. New York, Routledge.

Jones, P. (2015). Assessing the wider impacts of the Jubilee Line Extension in East London. *International Handbook on Transport and Development*. R. Hickman, M. Givoni, D. Bonilla and D. Banister. London, Edward Elgar (Forthcoming).

Markusen, A., Hall, P. and Glasmeier, A. (1986). *High Tech America: The What, How, Where and Why of the Sunrise Industries*. Boston, Allen and Unwin.

Markusen, A., Hall, P., Campbell, S. and Deitrick, S. (1991). *The Rise of the Gunbelt: The Military Remapping of Industrial America*. Oxford, Oxford University Press US.

McCurry, J. (2015). Japan's maglev train breaks world speed record with 600 km/h test run. *The Guardian*. London, Guardian News and Media Limited (Published on 21 April 2015, accessed on 22 June 2015).

Mitchell, R. B. and Rapkin, C. (1954). *Urban Traffic: A function of Land Use*. New York, Columbia University Press.

Pidd, H. (2015). Northern powercut: has the Tories' rail revamp reached the end of the line? *The Guardian*, Guardian News and Media Limited (published on 25 June 2015, accessed on 22 July 2015).
Sands, B. (1993). *The Development Effects of High-Speed Rail Stations and Implications for California*. California High Speed Rail Series, Institute of Urban and Regional Development University of California at Berkeley Berkeley.
Schumpeter, J. A. (1964). *Business Cycles: A Theoretical, Historical and Statistical Analysis of the Capitalist Process* (Abridged, with An Introduction, by Rendigs Fels). New York, McGraw-Hill.
Schumpeter, J. A. (1982 [1939]). *Business Cycles: A Theoretical, Historical, and Statistical Analysis of the Capitalist Process*. Philadelphia, Porcupine Press.
Teitz, M. (2014). "Peter Hall in California." *The Annals of Regional Science* 83(10): 438–439.
Thomson, J. M. (1977). *Great Cities and their Traffic*. London, Gollancz.
Veltz, P. (1996). *Mondialisation, Villes et Territoires: l'économie d'Archipel*. Paris, Presses Universitaires de France.
Weightman, G. and Humphries, S. (1983). *The making of modern London, 1815–1914*. London, Sidgwick & Jackson.

Chapter 5
The Strategic Planning Protagonist: Unveiling the Global Mega-City Region

Kathy Pain

5.1 Introduction

When Peter Hall first brought *The World Cities* to international attention in 1966, his analytical focus was to predate that of Scott's (2001) 'global city-region' by three decades. Drawing on Geddes' (1915) earlier treatise on the metamorphosis of 'world city' regional landscapes, Peter described the rise to prominence in the competitive international economy of certain physically extensive and supremely well-connected world city regions. In addition to London, New York, Tokyo (Saskia Sassen's three leading 1991 'global cities'), Paris and Moscow, he included the Randstad, Netherlands, and the German Rhine-Ruhr regions amongst his world cities, later referred to by Friedmann (1986) as a West European world city sub-system.

Decades ahead of other writers, Peter foresaw the importance of the role of new economy specialised global financial, business and professional services in the generation of internationally competitive urban regions. His interest in the evolution of this inherently global development phenomenon never waned but was transposed to the rapidly changing late twentieth century landscapes of East Asia (Hall 1999) and new millennium globalizing polycentric 'mega-city regions' in North West Europe (Hall/Pain 2006). His focus on the governance and planning challenges posed by the spread of 'global city work', city consumption patterns and individual travel by car, across enlarging functionally networked mega-city regions in Eastern and Western contexts is of ever growing world significance.

The North West European 'mega-city region' later identified by Peter Hall and his co-researchers in a 2003–2006 study, 'POLYNET: Sustainable Management of European Polycentric Mega-City Regions', was, he claimed, an unprecedented urban form (Hall/Pain 2006). But the urban processes underpinning this global

Kathy Pain, Professor, University of Reading, UK, Professor in Real Estate Development.
k.pain@reading.ac.uk.

R.D. Knowles and C. Rozenblat (eds.), *Sir Peter Hall: Pioneer in Regional Planning, Transport and Urban Geography*, SpringerBriefs on Pioneers in Science and Practice 52, DOI 10.1007/978-3-319-28056-1_5

development phenomenon have in reality been in evolution for centuries, albeit the changes with which they are associated have gathered great pace in recent decades and continue to do so. Furthermore it can be argued that the conceptual and analytical framing of the mega-city region today has been informed substantially by Peter's own thinking since the early 1960s. In consequence, the first and primary intention in writing this essay has been to research the emergence of the contemporary mega-city region as observed and understood by Peter whose analytical lens was focused obsessively on the spatial form and functioning of cities for more than half a century. Selected works from his voluminous list of contributions to the urban and regional studies literature are referred to.

Peter's passion for trains and railways, since his childhood, is well known. By the age of six when his family moved to Blackpool from Hampstead in North London, he had formed an obsessive interest in the London Underground, which led to his faultless knowledge of the geography of the entire network in later years. Ultimately trains, trams and modal transportation interchanges became the final focus of his active research. In 2009, he co-authored a report for the then UK Secretary of State for Transport on the future of railway stations and he subsequently developed major European studies promoting new transport technologies for five regions of North West Europe: 'SINTROPHER—Sustainable Integrated Tram-based Transport Options for Peripheral European Regions', and later, 'SYNAPTIC—Synergy of New Advanced Public Transport Solutions Improving Connectivity in North-West Europe'.

Working with Peter intensively as his Co-Director on the POLYNET study in the preceding years, I was struck by his enduring leviathon energy both as a student of rail networks and as a rail traveller. But looking back over events in the process of writing this essay, led me to speculate that his later very specific research focus on the development of European advanced public transportation systems was motivated not only by his lifelong attraction to trains and railways but also by fresh understanding of mega-city region processes and sustainability challenges that had come from the POLYNET study. A second intention in the essay therefore, is to trace how a critical relationship between Peter's two great passions, cities on the one hand and railways on the other, ultimately allowed him to inform a new, spatial vision for more sustainable mega-city regions.

But my familiarity with Peter's work dates from much earlier in his career when he was my Ph.D. supervisor in the Department of Geography at the University of Reading in 1973, just a decade after the publication of his own Ph.D. Some of the books and articles I have referred to in writing this essay have been in my personal collection (still marked with my study notes) since 1969 when I began by studying urban planning prior to my Ph.D. research with Peter. Over time, Peter's work has come to be widely seen as primarily focused on planning but in the 1970s he was highly active in contributing to critical debates in urban economic geography. His interests and knowledge were broad and interdisciplinary, going far beyond the Royal Town Planning Institute (RTPI) syllabus that I had studied as an undergraduate. Thus a third intention is to draw attention to the part played by the breadth of Peter's disciplinary positioning, first and foremost in geography, that allowed him to contribute critical insights to planning practice.

5.2 The Champion of Regional Planning

In the September 2014 issue of *The Planner* (the professional planning journal to which Peter had contributed frequently for much of his career) in a tribute to his work following his death aged 82 on July 30th of that year, Martin Read asked "does today's planning profession provide the right environment to attract and engage future Peter Halls?" (Read 2014: 5). Having worked with Peter close to both the beginning and the end of his career, I think that Read was asking the wrong question. My sense is that the reason that the planning profession was able to benefit greatly from Peter's insights was precisely because he was not himself trained as a practising planner. As a geographer with a doctorate on London's 1851–1939 historical-economic industrial location geography (Hall 1962), his pursuit of knowledge and understanding had been unconstrained by the requirements imposed by a practitioner's professional accreditation framework. Based on my experience of working with Peter, I would conjecture that it was his 'outsider' perspective as a non-practitioner and his primary focus on understanding cities that led the planning profession to be so attracted to and engaged with Peter. From 1992, he was a Professor in the Bartlett School of Planning at University College London. The knighthood for his services to the Town & Country Planning Association awarded in 1998 and the RTPI Gold Medal awarded to Peter in 2003 were recognition of his contribution to the profession.

Peter Hall's interest in the historical evolution of cities, and 'world cities' in particular, and their spatial form, always had obvious relevance for planning. Peter's supervision of my doctorate on housing markets and urban policy in the early 1970s demonstrated his interest in the critical evaluation of urban governance and planning practices. Following my Ph.D. studies with Peter, he chaired the new School of Planning Studies at the University of Reading that he had helped to launch. Titles of his book publications at that time illustrate his contribution to the state of knowledge on 'how to do' planning better, for example, *Theory and Practice of Regional Planning* (1970) and *Urban and Regional Planning* (1975). In *London 2001* (1989) Peter predicted the diverse consequences of the demise of strategic regional planning and the abolition of the Greater London Council in 1986, urban congestion, economic collapse, and social deprivation which he later documented in *London Voices, London Lives* (2007). He has been a longstanding champion of planning but generally from the standpoint of a 'critical friend' who wanted to make planning more strategic and effective in order to create better cities and better lives.

As early as 1963 in *London 2000*, Peter had argued for a new scale of strategic planning for London and its surrounding region to allow London to grow and to spread. This led to important South East studies and plans that Peter contributed to, for London's development as a multi-centred region. When these plans were later scrapped in 1983 by a Conservative Government, Peter's dismay was self-evident from his press commentaries. However it was in the two-volume book, *The Containment of Urban England* (Hall et al. 1973) that his deep concerns about the British planning system in the early 1970s are most apparent. With his co-authors, Peter sought to establish that by seeking to protect the countryside by enclosing London and other English cities with 'green belts', places for work and places for people had been separated from each other and, at the same time, land prices had risen. It was his long-held view that 'new towns' set in the countryside surrounding London, modelled on Howard's (1902) 'Garden City' or more accurately 'Social City' model, were a more sustainable arrangement for urban living and working because they created self-contained communities that did not require the need for commuting. As it was to later transpire, this analysis was faulty as demonstrated by the case of Milton Keynes in South East England (Pain/Hall 2006) though, curiously, ten years earlier in *London 2000* and in *The World Cities* (1966), Peter had foreseen the socio-economic changes that would lead to intense cross-commuting and leisure travel that would eventually give rise to the unsustainable living patterns (travel by car) in multi-centre or 'polycentric' regions worldwide. How this came about will be returned to in due course.

Peter later recalled that when he had served on the South East Regional Economic Council in 1967 helping to produce *The Strategy for the South East*, the document published (South East Economic Planning Council 1967, 1970) had 'caused a storm, not least because the chartered planners thought we were stealing their jobs'. And it is clear from his numerous periodical commentaries published at the time in for example, *Town & Country Planning* and *The Planner* and *Regional Studies*, that Peter lamented what he saw as the impoverishment and finally the abandonment of regional planning in England that was to follow (see also Tewdwr-Jones et al. 2014; Hall 2014a: 114–117, 2014b: 317–325). His strongly held views on the subject were repeated in many of his publications in the late 1980s such as 'The Anatomy of Job Creation: Nations, Regions and Cities' in the 1960s and 1970s (Hall 1987: 95–106) and in *London 2001* (1989) which challenged established regional planning thinking. In addition to his critique of the politically orchestrated demise of strategic planning, in *Good Cities, Better Lives* (Hall 2013a) he expressed his view that "…British planners have lost the art of urbanism" (p. 306).

5.3 From the Unsocial City to Regional Zones of Hope

As a politically engaged urban geographer, Peter had the ability to see beyond the limited sights imposed on spatial planning by its political context. His work has become associated in the minds of some critical geographers with the built environment and traditional 'physical' planning yet, ironically, his Sixth JR James

Memorial Lecture in 1986, the 'Unsocial City' (*From the Unsocial City to the Social City* reproduced in *The Planner*, 1986: 17–24) was critiqued by urban sociologist Ray Pahl for looking to "economic theories to explain social facts" (Pahl 1986: 10).

This was a time when the major geographical consequences of industrial decline and structural unemployment experienced since the 1970s in British cities that had once been manufacturing workshops of the world became evident. Critical contributions to geo-political debate in the 1980s by authors such as Massey (1983) and Peter drew attention to the scale of the challenge to spatially rebalance the economy of England that still persists today. The JR James Memorial Lecture demonstrates the ability Peter then had to introduce insights from urban geography and eruditely critique and inform urban policy. Going against the grain of 'area planning' policy wisdom of the time, he specified the real causes of the urban problem thus:

> …we are again obsessed by an inner-city crisis. But the real problem is not an urban one at all: it is a bundle of economic and social problems that happen to impact with especial force in some parts of cities. These problems are very deep and intractable; they have their roots in the development of our economies and of our societies – not just herein Britain, but also in other western countries – over the last 30 years (1986: 17).

His observation that contemporary economy and society and the relationships between them, have "urban expression" (1986: 17) also signaled the need to look beyond city metropolitan boundaries to find urban planning solutions.

In 'The People: Where will they go?' (Hall 1985: 3–12; Breheny/Hall 1996) Peter had called for planning to address the problems of "a massively unsocial city"—specifically, London's major housing deficit. And in the JR James Memorial Lecture he claimed that new towns offered "attractive terms for groups of the [urban] disadvantaged to move" (Hall 1986: 23). He identified Japanese style 'Technopolises' or "new towns deliberately constructed as the homes of the new industries" (1986: 23) as a means to secure the 'genius' of Howard's Social City of some ninety years before. Peter's Social City vision for late twentieth century England, was the creation of one or more clusters of "information technology, space exploration and exploitation, and biotechnology… built by a combination of public and private capital" (1986: 23). These clusters would be interlinked by 'corridors', for example the M11 motorway extended to Hackney (daring to breach the "green belt") connecting deprived inner-city areas and the new Technopolises and "progressively achieve a dispersal of the ghettoes of disadvantage into the new [city-region] zones of hope" (1986: 24).

His call for radical rethinking of regional planning policy was reiterated in the 1988 article 'The Industrial Revolution in Reverse' (Hall 1988: 15–19). Here Peter used the example of the "Golden Belt" of population increase in South East England to argue that "many kinds of economic and social activity may be carried out more efficiently, and often more pleasantly, outside the inner cities than within them" (1988: 18). This view was later to be expanded in 1998 in 'Sociable Cities: Legacy of Ebenezer Howard' (1998) with Colin Ward.

Peter's enthusiasm for the development of the Technopolis in Britain had been inspired by an academic post in the Department of City and Regional Planning at the University of California, Berkeley in the 1980s. This position brought him into close contact with 'Silicon Valley' and with sociologist Manuel Castells (author of *The Network Society,* 1996), leading to the co-authored book, *Technopoles of the World: The making of 21st Century Industrial Complexes* (1994). Fast-forwarding to 2014, Peter and colleagues would be presented with a special Wolfson Economics Prize Competition award for a proposal to take his Garden or Social City vision of the 1980s forward in a plan to develop twenty first century clusters of new and existing towns with names that evoke Patrick Geddes' call for civics, civic statesmanship and citizenship for "colossal city-groups" in Britain (Geddes 1915: 25)—for example, the 'City of Mercia' in the West Midlands, the 'City of Palatine' in Lancashire and the 'City of Kent' in the South East. For Peter, regional space was the place for the growth of new kinds of advanced work and a means of human escape from the 'Unsocial City'. He realized that to make this vision for social living a reality, progressive strategic planning was essential.

In an address to a Government organized South East conference on the SERPLAN (London and South East Regional Planning Conference) *A New Strategy for the South East* on 6th March 1991 (reproduced in *The Planner*, 22 March, 1991: 6–9), Peter Hall stated that he was "doubly sure" that the revival of Britain's cities "will provide the key to the revival of the regions". He expanded upon his vision of new transport 'corridors' holding the key to this regional revival (Hall 1991: 9) predicting that an 'M1' motorway link then under construction, and a new eastern corridor would "provide a major new line of force in the geography of Britain". He anticipated the shrinkage of time-distance travel with the roll-out of new motorways and high speed rail linking London to other major English cities such as Manchester and Birmingham (1991: 9). But he made it clear that this would require a transport corridors' 'strategy' to allow co-operation between district councils and the county councils who controlled the two main English growth corridors (1991: 9).

In the 1998 edition of *Sociable Cities* Peter proposed that to rebalance England, his clusters modelled on Howard should be developed along rail, tram and bus routes, public transport routes following the English West Coast, Midland and East Coast railway 'mainlines' and the high speed Channel Tunnel Rail Link (CTRL). It was an ambitious strategy that was echoed in the Office of the Deputy Prime Minister strategy for 'Sustainable Communities' (ODPM 2003, 2004) and the European Spatial Development Perspective which set out to address Europe-wide uneven development (EC 1999; Pain/Van Hamme 2015). But although developments such as the expansion of Ashford, Ebbsfleet and Bicester were eventually to be approved, Peter's aspirational vision for the implementation of a strategic development plan for the "Greater South East" was not to be realised within his lifetime.

5.4 The European Mega-City Region

The mega-city region identified and analysed in European research between 2003 and 2006, was inspired by Peter's longstanding study of London and South East England on the one hand and on the other hand by a fortuitous event that occurred in 2002.

As a then Senior Researcher in the Globalization and World Cities Research Network (GaWC) at Loughborough University I had invited Peter to chair a focus group meeting of senior government officials and business actors in a Corporation of London/Economic and Social Research Council (ESRC) study of advanced business services clustering in the City of London. Immediately after that meeting, Peter proposed that we put our heads together to design a South East England 'mega-city-region' study. This would bring together Peter's experience functional urban region (FUR), otherwise known as functional urban area (FUA) analysis, based on commuting flows and GaWC experience of analysing the flows associated with city-based global advanced business services networks. The project would be based at the Institute of Community Studies in London where Peter was then the Director from 2001 to 2004.

A first proposal submitted by us to the UK funding council, the ESRC, was rejected: 'Deconstructing the Mega-City-Region: A case study of South-East England'. However there was enthusiastic interest in our research idea when I attended a European Commission INTERREG IIIB Programme 'Spark-off' project networking event in Paris. The proposal we subsequently submitted to the Programme was successful, winning us EUR 2.4 million European Union and member state funding to undertake a three year study of eight regions of North West Europe: 'POLYNET: Sustainable Management of European Polycentric Mega-City Regions'. The project team comprised researchers from GaWC at Loughborough and from each of the eight mega-city regions: South East England; The Randstad, Netherlands; the German Rhine Ruhr and Rhine Main; Central Belgium; Paris; Northern Switzerland; and Dublin. A bonus was that members of the Netherlands and Rhine-Main research teams brought to the study experience of 'polycentric urban region' (PURs) analysis which had also informed the City of London business cluster study.

The term 'megacity' had long been used to refer to very large cities for example in research by Janice Perlman's *Mega-Cities Project* (www.megacitiesproject.org) and especially in describing mushrooming cities in the global South. However the urban phenomenon that had long held Peter's interest was not simply about size, it was an interconnected regional entity. This phenomenon had been proposed in 1961 by Jean Gottman in his notion of the 'Megalopolis' located the urbanized North Eastern Seaboard of the United States of America. And at the turn of the century in the United States and in Eastern Asia, a variety of terminologies were beginning to be used to describe very large regions comprising multiple urban centres and interconnected by traffic flows.

Gottmann's *Megalopolis* (1961) had described the process by which the towns and cities proximate to world city New York, were becoming functionally interconnected to a degree that together they could be seen as being effectively one urbanised entity spanning a large geographical area. When Peter observed the apparently similar process that was occurring in the rapidly urbanizing Pearl River Delta and Yangtze River Delta coastal areas of China, he had seen an analogy with South East England. But by introducing GaWC analysis to the study of such regions in the POLYNET study it became possible to gain understanding of how the global connections of '*The World Cities*' (1966) at the heart of some regions might spread to a wider city-region scale. The project idea recognized that urban processes are subject to metamorphosis and can also spill out over extensive landscapes that cross administrative city and regional boundaries. Thus, the mega-city region studied in North West Europe is described in the Polycentric Metropolis as, …a series of anything between ten and 50 cities and towns, physically separate but functionally networked, clustered around one or more larger central cities, and drawing enormous economic strength from a new functional division of labour. These places exist both as separate entities, in which most residents work locally and most workers are local residents, and as parts of a wider functional urban region (FUR) connected by dense flows of people and information carried along motorways, high-speed rail lines and telecommunications cables: the 'space of flows' (Castells 1996: 376–428) with major implications for sustainable development (Blowers/Pain 1999). It is no exaggeration to say that this is the emerging urban form at the start of the 21st Century (Hall/Pain 2006: 3).

5.5 The Global Mega-City Region Unveiled

So the megacity region studied in Europe was not simply a highly interconnected urbanised space, it was a potential landscape of the competitive international economy. Decades earlier Peter Hall had foreseen the importance of the role of new economy specialised global financial, business and professional services in the generation of internationally competitive urban regions. The rapidly changing new millennium regional landscapes of world cities in the US, Europe and East Asia were not only distinguished by intensifying movement into, out of, and within them, but by the apparent spreading out, spatially, of global city work and lifestyles.

Functionally networked both globally and locally, in an integrating world economy, these extensive mega-city regions were becoming globalized phenomena. The mega-city region concept used in POLYNET therefore resonated with Scott's (2001) global city-region however a notable distinction was the inclusion of Hall's 1966 'world city regions', the Randstad and the Rhine-Ruhr, and other urban regions not given prominence in the 'global cities' literature (Sassen 2001) in addition to London and Paris.

Global or globalizing mega-city regions are networked by complex flows of people, goods and information. The research therefore attempted ambitiously to map the complex flows that characterize the eight mega-city regions not only of people such as in physical commuting and business travel, but of ideas that travel through business networks in the form of face-to-face and virtual communications of all kinds (Pain/Hall 2008). The results are reported in a book, *The Polycentric Metropolis: Learning from Mega-City Regions in Europe* published by Earthscan (Hall/Pain 2006) and in numerous articles written since the completion of the study.

5.6 Two Types of Region—Two Types of Mega-City Region?

Peter's past thinking was to prove highly relevant to issues of regional definition and analysis. In 1970, in Chap. 2 of his text book *Theory and Practice of Regional Planning*, Peter had considered the ideas of the French writer Boudeville (1966) who defined two different types of region. This distinction was to have strong resonance with a critical difference between two types of mega-city region identified in the POLYNET study (Hall/Pain 2006; Pain et al. 2006; Taylor/Pain 2007; Pain 2008).

Boudeville's first regional type was called a 'homogenous' or "statistically uniform" or "static region" which exhibited statistical uniformity within a certain range based on a particular criterion, for example, population density. A "contiguity constraint" was imposed, in other words, for a homogenous region to be statistically uniform, its component areas must be homogenous and contiguous with each other for the same criterion (1970: 14–15). Principal components analysis was then used to define "real natural regions" on the basis of statistical variations and areas of similarity for more than one criterion (1970: 15). His second regional type was called a 'nodal' or "dynamic flow region". This type of region reflected patterns of movement in geographical space, for example commuting.

In 1970, Peter had noted the potential for these regional 'types' to inform analysis. For example he referred to the potential use of data on periodic or regular trips such as "business men travelling on their regular routes" (1970: 15) as applied, together with commuting and other flow data in the POLYNET study. He also foresaw that flow analysis could be applied to larger scale commodity movements which have been the focus of subsequent US 'megaregion' studies, for example Catherine Ross's research on megaregion freight and commodity flows and their economic and environmental impacts (Ross 2009; http://www.coa.gatech.edu/people/catherine-ross).

A characteristic of Boudeville's flow regions was their 'nodal' or "hierarchic structure" whereby "flows out of smaller centres converge and coalesce in bigger ones; in turn, the bigger flows from these bigger centres, together with flows of smaller centres, may flow into a bigger centre still" (1970: 15–16). This conceptualization of the network architecture of a flow region was later illustrated by patterns of daily commuting in the POLYNET study which referred to the contiguity of daily travel to work areas, or FURs, to describe the functional mega-city region (see Figs. 5.1, 5.2 and 5.3).

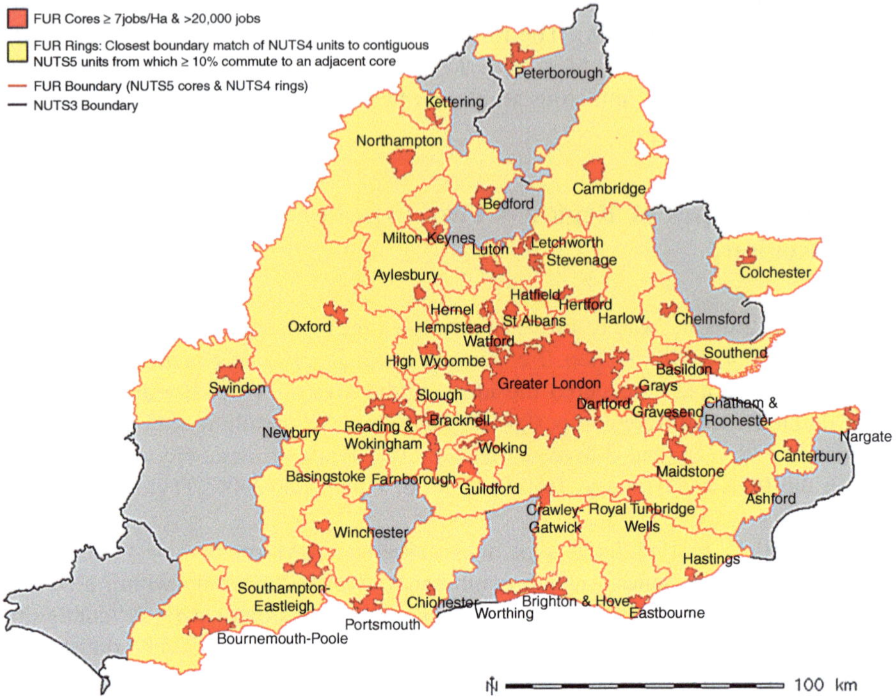

Fig. 5.1 South East England Mega-City Region: Constituent FURs. *Source* Hall/Green (2005: 4)

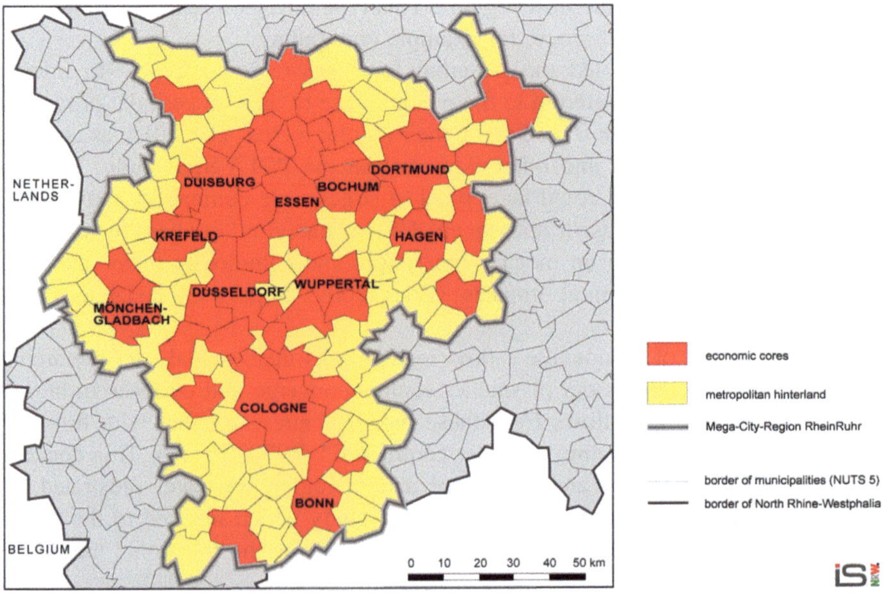

Fig. 5.2 Rhine-Ruhr Mega-City Region: Constituent FURs. *Source* Knapp et al. (2005: 3)

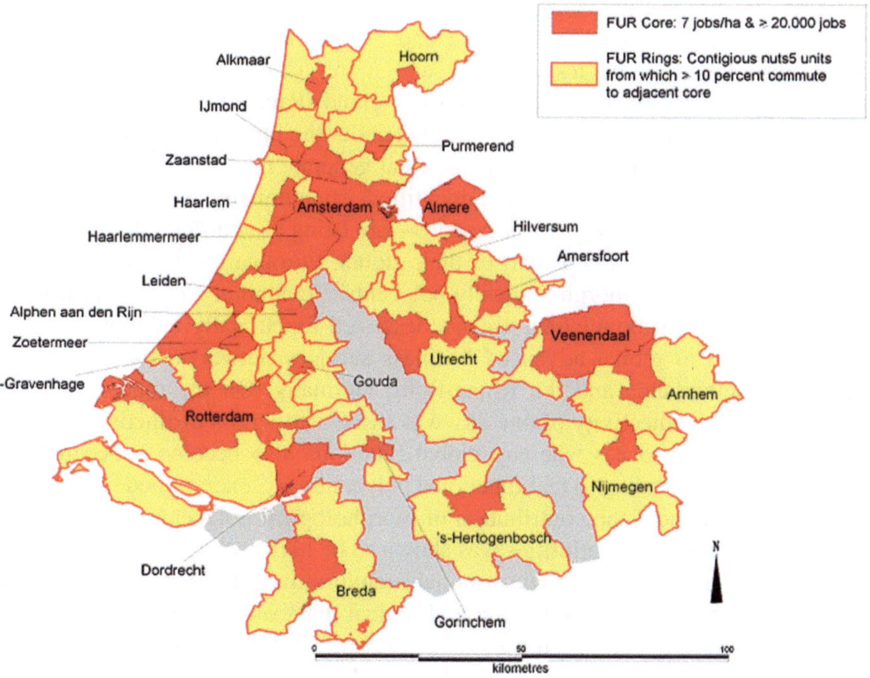

Fig. 5.3 The Randstad Mega-City Region: Constituent FURs. *Source* Werff et al. (2005: 3)

Making reference to the 1930s work of Walter Christaller, Peter saw in 1970 that the systems of flow described by Boudeville could potentially also correspond, "…in a systematic way in many cases to levels of economic activity which are associated with the centres or nodes themselves. They form, therefore, a sensitive guide to the functioning of the local economy." (1970: 16). He referred specifically to the economic significance of flows of population engaged in 'non-primary' industries that were later to be the critical focus of the POLYNET research.

Although Peter never referred directly to Boudeville's work, or to his *Theory and Practice of Regional Planning* analysis when we designed the POLYNET project, the two regional types that Boudeville had described can be seen in retrospect to have common reference points with mega-city region distinctions that were later identified. Boudeville's flow region corresponds to what was termed in POLYNET a 'functionally polycentric' mega-city region. This mega-city region type is interconnected by networks and flows at multiple global to local scales. Significantly, the more global the networks and the flows of a city are, the more global is its mega-city region. Boudeville's homogenous region corresponds to what was termed a 'morphologically polycentric' mega-city region in POLYNET, for example the Rhine-Ruhr and the Randstad. This mega-city regional type has a profile of statistical similarity, for example the population size of the urban centres in a mega-city region, giving an impression of spatially balanced development, for example, the Rhine-Ruhr, Germany, and the Randstad, Netherlands, compared to South East England (see Figs. 5.1, 5.2 and 5.3). Notably, this type of mega-city region was found to be less 'global' and more 'static' than the 'functionally polycentric' mega-city region, South East England (Pain 2006, 2008, 2011b).

Of significance is what the study revealed about the importance of different roles and functions of the global or world cities that generate mega-city regions. The two different types of mega-city region identified—functionally polycentric regions, versus morphologically polycentric regions—are being used differently by advanced business services networks as part of their global, European and national strategy. Regions that are more morphologically polycentric, containing an even distribution of cities of a similar population size, are less well functionally integrated and less well connected to global advanced business services networks, for example the Randstad, Netherlands and the Rhine-Ruhr, Germany. On the other hand, high global network connectivity of one city was associated with a distinctive mega-city region expansion process whereby global functions also become concentrated in proximate centres leading to real functional polycentricity, exemplified in South East England. Furthermore the contribution of morphological polycentricity to regional sustainable development priorities was shown to be highly questionable. The rising level of cross-cutting travel has unsustainable environmental effects. In the Randstad for example, the urban concentration and critical mass required to support efficient public transportation was missing and planning policies for the containment of urban areas were found to be failing. Meanwhile the Randstad was shown not to have a balanced economic structure even though it exhibits morphological polycentricity. Reflecting on Boudeville's terminologies cited by Peter in 1970, the regional distinction made in POLYNET can therefore usefully in retrospect be specified as: I—A dynamic flow/nodal mega-city region (for example, South East England) and II—A static/homogenous mega-city region (for example, the Randstad, Netherlands).

5.7 Two Mega-City Region Processes

A dynamic flow/nodal versus static/homogenous conceptualization of the two mega-city region types has resonance with later reflection on the POLYNET results which drew on Jacobs (1969) and Castells (1996) to inform a programme of US 'megaregion' research. In Taylor/Pain (2007), a distinction is drawn between two mega-city region processes 'Process A—Mega-city region expansion' and 'Process B—Construction of mega-regions of proximate cities'.

Process A—'Mega-city region expansion' is that identified in South East England where mega-city region expansion is occurring in spite of the primacy of London. London's expansion is enveloping previously proximate cities and towns (for example Reading, Berkshire) that exhibit multi-sector advanced business services clustering and a strong presence of international networks and connectivity. This process is one of a diffusion of urban functions envisaged by Jacobs, but at a much larger geographical scale creating a functionally polycentric 'global mega-city region' structure with multiple connections to the world economy (Pain et al. 2006).

Process B—'Construction of mega-regions of proximate cities' is identified in mega-city regions exhibiting varying degrees of lesser primacy than London. Of note, these regions showed sectoral specialisation between their cities (for example banking/advertising in Amsterdam, architecture/logistics in Rotterdam, management consultancy in Utrecht-Amersfoort) and a functional differentiation in terms of the scale of their business network connections with one leading city (for example, Amsterdam and Düsseldorf) which are dominant for global connectivities. The result is a primate centre functionally surrounded by a homogenous or static region that is functionally disconnected from the primate city. As Taylor and Pain put it, "Thus rather than creating a Jacobsean city-region, process B is about constructing mega-regions of proximate cities [or] an urban region with holes in it" (2007).

Returning to 1970, in Peter's discussion of the difficulties arising from the application of Boudeville's regional distinction in British situations, he illustrated the kernel of future debates in spatial analysis in relation to questions of scale and governance. In considering the question of scale, he notes that,

> …whole sets of regions can in fact be observed at different levels … there is a nesting principle, so to speak, in which larger numbers of smaller regions fit inside smaller numbers of larger regions … therefore the pattern of nodal regions which one produces in any one area is not a single pattern of regions. There is no unique set of nodal regions; there are a number of possible sets of nodal regions depending on the level of activity one is specifying. Then again, we have just been talking about a single function … but if we look at different functions the set of regions and their boundaries may in fact vary (Hall 1970: 16).

Furthermore he saw that,

> …there is a problem greater than this. It is that by definition, there is no necessary correspondence between homogenous regions and nodal regions, or between sets of homogenous regions and sets of nodal regions. Indeed if one thinks about this for a little while, there is a good reason why this should not be. The homogenous regions may be defined as homogenous according to any number of statistical characteristics … there may be a relationship between homogenous regions and nodal regions [where common causes exist], but on the other hand, there may not be (1970: 17).

He therefore drew attention to the contested nature of 'the region', its scale and its definition, that would come to be subjects of urgent critical debate in the 2000s (Harrison/Pain 2012; Pain/Van Hamme 2014). The POLYNET research revealed polycentrism to be a scale-sensitive phenomenon. The globalization of advanced business services activity is generating functional polycentricities at global to local scales, including the mega-city region scale and an emerging transnational, 'metacity' scale (Pain 2010). This finding has important implications for policy and planning as illustrated in England by Government priorities introduced in 2014 for a 'Northern Powerhouse' comprising the Liverpool, Manchester, Leeds, Sheffield, Hull and Newcastle city regions, which is effectively one multi-centred mega-city region. An advantage of understanding mega-city regions as a process is that overlapping scales and the likely reality that different processes coincide in the same space can be taken into account. The issues arising for mega-city regions planning and governance will be considered shortly.

5.8 Mega-City Region Zones of Hope—How Sustainable?

Returning briefly to the year 1963, in *London 2000* Peter had foretold part of this future mega-city region reality in his vision of London's evolution into a multi-centred region. It was here that he identified the coming of a "mega-city-region" that would be 120 miles wide and contain no less than 28 new towns. A fictional family, the Dumills, were living in a future compact new town to be called Hamstreet near Ashford in Kent built in 1973. The story was lighthearted but, importantly, it predicted a South East England society that was to prove incredibly accurate. Sophisticated city consumption behaviours had spread across the London region.

Hamstreet had been designed to have a target population of 95,000 and that target was under revision to cater for next generation families. 70,000 of that target population would be able to walk to the town centre shopping area. But despite its design as a self-contained community, three of the four members of the Dumill family do not work in Hamstreet. As an ex-city dweller, Mr. Dumill commutes daily to work in London. Moreover, while Hamstreet's fictional planners were alarmed at growing violation of the new town enclosed community ideal, they came to accept it as an inevitable consequence of increasing personal mobility: Hamstreeters "thought as little of going 25 miles to work as of driving 50 miles in the evening to see friends, or going over to Brittany for a weekend" (London 2000 (1963), reproduced in Hall (2000: 44) The Dumills—Londoners 2000, *Town & Country Planning*).

Furthermore, for the Londoners of 2000, "distance is no longer an object. They learned to do without cars for much of their working day, but for living their own lives the car has become part of themselves. Just as they no longer can work in isolated communities, so they can no longer form their friendships and their social lives within bounds of space. For them Hammersmith to Dover has no more significance than would Hammersmith to Hampstead in 1960" (2000: 60). The scenario of the spread of city consumption patterns and individual travel by car across an enlarging functionally networked mega-city region was also discussed in relation to 'The World Cities' specifically by Peter in 1966, where he also foresaw the governance and planning challenges that would arise in South East England four decades later.

As a reverse commuting resident of West London driving his Citroen Deux Chevaux car some 40 miles daily down the M4 to work at the University of Reading in the 1970s, Peter would not have been unfamiliar with the reality of complex patterns of mobility that would be generated by the residents of a multi-centred region with a world city at its centre. Yet, in the year 2000, reflecting on how his

1963 vision had turned out, he spoke of his failure to spot the development of Milton Keynes but did not discuss the implication of 'The Dumills' lifestyle, i.e. that new towns scattered across a regional landscape close to London might be less self-contained than Howard could ever have predicted in writing *Tomorrow* in 1898 (reprinted in 1902 as *Garden Cities of Tomorrow*). Peter commented optimistically in 'The 1963 view from 2000' (Hall 2000) that Many things in this prediction have come out nearly right … others have not, but the reason is interesting and saddening: the system has been too slow to anticipate and to react, so that events I predicted to happen 30 years ago may happen—with luck—in the next ten. Hamstreet did not become a new town (a pity), because it was pure sustainable development 2000-style… (2000: 46).

His continued vision of a polycentric Howard-style London region populated by more new towns by the year 2000 imitated the Dutch Randstad development model. But this was shortly prior to the POLYNET study that was to identify the unsustainable environmental impacts associated with morphological polycentricity due to intense cross-cutting personal mobility by car (see Fig. 5.4).

But, as will be seen, by the year 2014 at the end of Peter's life, he began to see how Howard's Social City might be applied more sustainably in contemporary European mega-city regions.

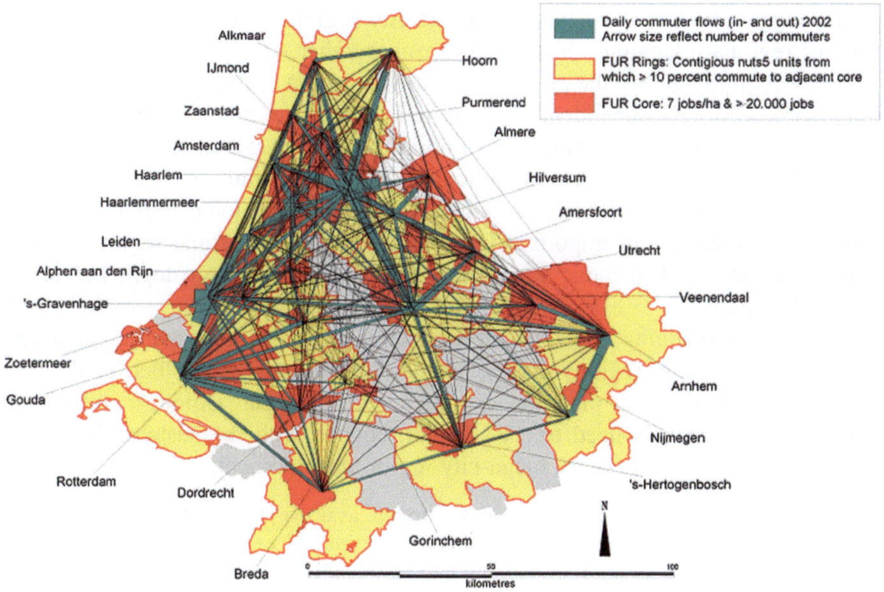

Fig. 5.4 The Randstad Mega-City Region: Commuting, 2002. *Source* Werff et al. (2005: 14) and Hall/Pain (2006: 38)

5.9 Two Types of Regional Planning: Two Types of Mega-City Region Planning?

Returning once more to *Theory and Practice of Regional Planning* (1970), in what ways can Peter's consideration of a distinction between two types of British regional governance inform the mega-city region planning challenges of today?

Drawing heavily on Self (1967), he identified two types of regional planning, the first deriving from an economics or regional science discipline approach and the second from ecology, design and land economics. The first type of regional planning referred to as 'national/regional' concerned national investment allocation planning in which policy aims to steer economic growth spatially by introducing measures by using development incentives and steering resource allocation. Peter argued that this type of regional planning was the one that had become the dominant European model (Pain 2011a). The second type of regional planning referred to as "regional/local" concerned the planning of the city and city-region. Here Peter pointed to the blend of resource, land use and design policies that relate to this scale which he referred to as an intermixing of "non-physical planning and physical planning" (1970: 23).

Peter made the interesting point that these two types of British regional planning had historically rarely appeared to "run together" and that when they had done so it could be argued that "the marriage was a forced one" (1970: 24). The POLYNET results suggest that the two types of mega-city region demand that there be two types of regional planning which need to run together.

He went on to identify the problem that the definition of regional and city-region boundaries is entirely dependent on the criteria used. Thus "the regional scale" will be variable and moreover "whole sets of regions can be observed at different levels" (p. 16). He identifies "the greater difficulty", "that by definition, there is no necessary correspondence between homogenous regions and nodal regions" (p. 17). These issues clearly had important implications for planning and regional governance structures and agendas which had not been explored in depth before the POLYNET study.

Coordinated planning was found to be compromised by governance challenges in all of the mega-city regions studied. Halbert and co-authors (Halbert et al. 2006: 206–218) emphasise that planning approaches at European Union, member state and sub-national levels need to, but do not, engage with the sustainability challenges posed by contemporary mega-city region development,

> The vague and ill-defined concept of polycentricity, as a morphological state, in current documents is hard for policy-makers to apply and, even if it were possible, this form of polycentricity at a regional scale has no association with increased economic competitiveness, environmental sustainability or territorial equity (p. 214).

The authors go on to give examples of the lack of appropriate governance structures in mega-city regions that are identified in the study as having, "a prevalent soft and shifting functional geography" (p. 208):

In the Zürich-Basel case Swiss policymakers,

> …remain preoccupied with small-scale territorial governance issues and do not adapt the political framework to emerging issues that will be – and even already are – crucial for Swiss integration into the global economy." And the boundaries of its institutional structures are problematic with an "underlying incapacity to define and implement policies at the MCR scale" (p. 208).

In the case of Paris,

> …the fact that the 1994–1999 *Contrat de Plan Interregional du Bassin Parisien* (CPIBP) was not prolonged in a successive period, like all other *Regional Contrats de Plan* (2000–2006), was a clear sign in the Paris Region of the inability of various policymakers (national and regional) to work hand-in-hand to develop coherent policies at the MCR scale (p. 208).

In South East England, in spite of growing recognition that city-region boundaries are 'fuzzy',

> …the extent and significance of inter-urban functional linkages is not reflected in either spatial policy advice or institutional and administrative arrangements … the structures, powers and resources required to manage intensive MCR processes are currently lacking, resulting in an institutional 'thinness' in the area of urban interaction where significant changes are occurring (p. 208).

Since advanced business services networks operate in and connect multiple cities functionally, territorial competition between cities is unhelpful especially since cities do not fit neatly into 'regional boxes'. As Peter pointed out in 1970, "Whole sets of city regions can be defined depending on the criterion you choose for the nodal region" and the space of flows in the contemporary global service economy is agile, the specification of mega-city region boundaries has ultimately proved a fruitless and dangerous exercise (see Harrison/Pain 2012). Taylor/Pain (2007) conclude that,

> The reason why this identification of different processes is important is because policy should be built upon process: two processes require two different policies. The main policy weakness identified in Europe has been a failure to conceptualise spatial relations in this way, hence the need to support dynamic and fluid mega-city regions has not been addressed … the materialist approach adopted in the research reported highlights the need to obtain evidence for the veracity of regional concepts in the work carried on in cities: planning should not be carried out separate from the practice of current economic actors (firms) that use cities.

5.10 The Regional Planning Protagonist—Final Reflections

In introducing this chapter, I made reference to three intentions that have proved to be interlinked: First, to research the emergence of the contemporary mega-city region as observed and understood by Peter whose analytical lens was focused obsessively on the spatial form and functioning of cities for more than half a century; second, to trace how a critical relationship between Peter's two great passions, cities on the one hand and railways on the other, ultimately allowed him to inform a new, spatial vision for more sustainable mega-city regions; and third, to draw attention to the part played by the breadth of Peter's disciplinary positioning, first and foremost in geography, that allowed him to contribute critical insights to planning practice.

The short trawl through a selection of Peter's publications has highlighted how his early interest in the historical-economic geography of London during his Ph.D., was to evolve over time, introducing an interdisciplinary spatial perspective to urban policy and planning debate. His early ideas for the roll-out of Social City new towns across South East England can be viewed in retrospect as having been destined (had they been realised) to contribute to a regional development pattern in North West Europe that is now recognised as generating unsustainable intense, cross-cutting physical flows of people by car.

Howard's Social City model, upon which Peter's new towns vision was based, was formulated years before the advent of mass car ownership. The ideal Garden City of 1902 was predicated on a totally different model of mobility needs, expectations and means to that of today, i.e. the use of feet, and even horses, instead of cars. Thus the sustainability of Howard's model clearly requires critical re-evaluation and re-working (Adams et al. 2015). Moreover, although 'urban containment' had become a planning objective when *London 2000* was written, this was years before the societal changes that would present new challenges to containment policy and before widespread recognition of linked climate change and environmental sustainability imperatives.

Ultimately, Peter's two great passions were to come together in his pursuit of a model for the future sustainable development of mature polycentric mega-city regions. He came to realise the need to review his regional vision post-POLYNET, and in the same year that *Good Cities, Better Lives* (2013a) was published, he addressed the crucial question that had eluded attention in his earlier work in 'Refreshing the parts that other transport cannot reach' (Hall 2013b):

> But the big question still remains: can tram-trains and BRT services perform the miracle of extending good public transport into wider and wider peripheral areas? The answer is probably not… (2013b: 130).
>
> …the area west of London is turning into a complex mega-city region of its own; but as well as being dependent increasingly on long trips to London, it is also developing what one could call a life of its own. The growth in these long 'inter-unit' journeys is itself very worrying (2013b: 131).

In *Sociable Cities* ... (Hall/Ward 2014), recognising the increasing urgency of the challenge to address the present and future housing needs of London's growing population beyond its boundaries, post-SINTROPHER, Peter saw the virtue of the European model for the development of transit-linked 'urban extensions' within short distances of existing town centres. Because UK densities of dispersed smaller urban centres potentially suitable for expansion would prohibit tram links, he proposed rapid bus transit to connect the new settlements to high-speed train stations.

However to achieve such a vision in England, a strategic plan to succeed the ODPM (2003, 2004) strategy for 'Sustainable Communities' would need to be reinstated and local communities and authorities would need to cooperate to deliver it. Peter's long held view was that Britain was in need of a radical new interventionist approach to planning and governance at the two levels he had discussed in *Theory & Practice of Regional Planning* in 1970: first strong state engagement with national/regional level priorities and central policies for infrastructure investment throughout the country and, second, devolution of financial levers to empower regional/local level planning and governance for real functional city-regions.

Writing in 1975, Peter had mourned the loss of joined up regional planning after the demise of the 1950s Ministry of Town and Country Planning regional offices in each of Britain's main provincial cities,

> …at this point the idea of coordinating the various local plans seems to have been more or less abandoned. The almost inevitable result was that the various local planning authorities, left to their own devices, pursued a defensive and negative policy (Hall 1975: 123–124).

And following the demise of the regional planning system that had eventually been introduced in England by a Labour Government, there are no longer any real vehicles for the development of long term, coordinated spatial strategies and co-operation between Local Authorities, Local Enterprise Partnerships (LEPs) and cities with combined authorities and City Mayors across England. Peter was a strong advocate for strategic planning however it can be argued that it was his disciplinary positioning as a planning 'outsider' that allowed him to make his most important contribution to good cities, better lives and better planning, a fearlessness in remorselessly questioning the status quo.

Given his longstanding contributions to debate on a strategic planning vision for London, it is no surprise that he came to study the global mega-city region in South East England that had been emerging throughout his career in what proved to be its final decade. Looking back, the mega-city region he identified in England and researched with colleagues across North West Europe seems to have been a natural development from everything in Peter's career that had gone before. Significantly, the culmination of that research has ongoing relevance for the sustainable management of urbanization worldwide.

References

Adams D., Larkham P. and Pain K. (2015) Viewpoint: Revaluating the place of urban planning history, *Town Planning Review*, 86(2), pp. 373–379.

Blowers A. and Pain K. (1999) The Unsustainable City? In *Understanding Cities: Unruly Cities? Order/Disorder*. Pile S., Brook C. and Mooney G. (eds.), London, Routledge, 247–298.

Boudeville J.R. (1966) *Problems of Regional Economic Planning*. Edinburgh, Edinburgh University Press.

Breheny M.J and Hall P. (eds.) (1996) *The people—where will they go?* National Report of the TCPA Regional Inquiry into Housing Need and Provision in England. London, Town and Country Planning Association.
Castells M. (1996) *The Information Age: Economy, Society and Culture, Vol. 1: The Rise of the Network Society*. Oxford, Blackwell.
Castells M. and Hall P. (1994) *Technopoles of the World: The Making of 21st Century Industrial Complexes*. London, Routledge.
EC (1999) *ESDP: European Spatial Development Perspective: Towards Balanced and Sustainable Development of the Territory of the European Union*. Brussels, European Commission.
Friedmann J. (1986) The world city hypothesis. *Development and Change*, 17, 69–83.
Geddes P. (1915) *Cities in Evolution. An introduction to the town planning movement and to the study of civics*. London, Williams.
Gottman J. (1961) *Megalopolis. The urbanized northeastern seaboard of the United States*. New York, The Twentieth Century Fund.
Halbert L., Pain K. and Thierstein A. (2006) European Polycentricity and Emerging Mega-city Regions: 'One size fits all' policy? *Built Environment*, 206–218.
Hall P. (1962) *The Industries of London since 1861*. London, Hutchinson.
Hall P. (1963) *London 2000*. London, Faber and Faber.
Hall P. (1966) *The World Cities*. London, Weidenfeld and Nicholson.
Hall P. (1970) *Theory & Practice of Regional Planning*. London, Pemberton Books.
Hall P. (1975) *Urban and Regional Planning*. Harmondsworth, Penguin Books Ltd.
Hall P. (1985) The people: Where will they go? *The Planner*, 71(4), 3–12.
Hall P. (1986) From the Unsocial City to the Social City. *The Planner*, March, 17–24.
Hall P. (1987) The Anatomy of Job Creation: Nations, Regions and Cities in the 1960s and 1970s. *Regional Studies*, 21, 95–106.
Hall P. (1988) The Industrial Revolution in Reverse? *The Planner*, January, 15–19.
Hall P. (1989) London 2001. London, Unwin Hyman Ltd.
Hall P. (1991) A New Strategy for the South East, *The Planner*, 22nd March, 6–9.
Hall, P. (1999) Planning for the Mega-City: A New Eastern Asian Urban Form? In *East West Perspectives on 21st Century Urban Development: Sustainable Eastern and Western Cities in the New Millennium*. Brotchie J., Newton P., Hall P. and Dickey J. (eds.). Aldershot, Ashgate, 3–36.
Hall P. (2000) The view from 1963 and the 1963 view from 2000. *Town & Country Planning*, February 69, 44–46.
Hall P. (2007) *London Voices, London Lives: Tales from a Working Capital*. Bristol, Policy Press.
Hall P. (2013a) *Good Cities, Better Lives: How Europe Discovered the Lost Art of Urbanism*. Abingdon, Routledge.
Hall P. (2013b) Refreshing the parts that other transport cannot reach. *Town & Country Planning*, March, 82(3), 121–132.
Hall P. (2014a) We have a problem—so what to do? *Town & Country Planning*, 2014, 114–117.
Hall P. (2014b) Apologia pro vita sua. *Town & Country Planning*, August, 317–325.
Hall P. and Green N. (2005) *Polynet Action 1.1 Commuting & the Definition of Functional Urban Regions, South East England*. London, Institute of Community Studies/ The Young Foundation.
Hall P. and Pain K. (eds.) (2006) The Polycentric Metropolis: Learning from Mega-City Regions in Europe. London, Earthscan.
Hall P. and Ward C. (1998) *Sociable cities: Legacy of Ebenezer Howard*. Chichester, John Wiley & Sons.
Hall P. and Ward C. (2014) *Sociable cities: The 21st-Century Reinvention of the Garden City*. London, Routledge.
Hall P., Thomas R., Gracey H. and Drewett R. (1973) *The Containment of Urban England*. London, George Allen and Unwin.

Harrison J. and Pain K. (2012) *Global Cities, Volume IV: Planning and Governance of Cities in Globalization*. London, Routledge.

Howard E. (1902) *Garden Cities of Tomorrow*. London, Swann-Sonnenschein & Co Ltd, 141–151.

Jacobs J. (1969) *The Economy of Cities*. New York, Random House.

Knapp W., Scherhag D. and Schmitt P. (2005) *Polynet Action 1.1 Commuting & the Definition of Functional Urban Regions: Rhine-Ruhr*. London, Institute of Community Studies/ The Young Foundation.

Massey D. (1983) The Shape of Things to Come. *Marxism Today*, April, 18–27.

ODPM (2003) *Sustainable Communities: Building for the Future*. London, Office of the Deputy Prime Minister.

ODPM (2004) *Creating Sustainable Communities: Making it Happen the Northern Way*. London Office of the Deputy Prime Minister.

Pahl R. (1986) Ray Pahl on Hall's Social City. *The Planner*, April, 10.

Pain K. (2006) Policy Challenges of Functional Polycentricity in a Global Mega-City Region: South East England. *Built Environment*, 32(2), 194–205.

Pain K. (2008) Examining Core-Periphery Relationships in a Global Mega-City Region—The Case of London and South East England. *Regional Studies*, 42(8), 1161–1172.

Pain K. (2010) The Emerging Megacity in North Western Europe and Policy Ramifications for Sustainable Transport Networks. In *The emerging cross-border mega-city region and sustainable transportation*, KOTI-EWC 2010 Special Report, Mo, C., Kim, Y.H. (eds.), Hawaii/Korea, East-West Center/The Korea Transport Institute, 2–27.

Pain K. (2011a) 'New Worlds' for 'Old'? Twenty-First-Century Century Gateways and Corridors: Reflections on a European spatial perspective. *International Journal of Urban & Regional Research*, 35(6), 1154–74.

Pain K. (2011b) Spatial transformations of cities—Global City Region? In Mega City Region? *International Handbook of Globalization and World Cities*, Derudder, B., Hoyler, M., Taylor, P.J., Witlox, F. (eds.), Cheltenham, UK, Edward Elgar, 83–93.

Pain K. and Hall P. (2008) Informational Quantity versus Informational Quality: The perils of navigating the space of flows. *Regional Studies*, 42(8), 1065–1077.

Pain K. and Van Hamme G. (2014) *Changing Urban and Regional Relations in a Globalizing World: Europe as a Global Macro-Region*. Cheltenham, UK, Edward Elgar.

Pain K. and Van Hamme G. (2015) Europe as a Global Macro-Region: How is economic geography changing the Cohesion policy challenge? In *Regional Studies Association Conference, Challenges for the New Cohesion Policy in 2014–2020: An Academic and Policy Debate*. 04–06 February 2015, Riga, Latvia, Regional Studies Association.

Pain K., Hall P., Potts G. and Walker D. (2006) South East England: Global Constellation. In *The Polycentric Metropolis: Learning from Mega-City Regions in Europe*, Hall, P. and Pain, K. (eds.). London, Earthscan, 125–136.

Read M. (2014) Peter Hall—will we ever see his like again? *The Planner*, September, 5.

Ross C. (2009) *Planning for Global Competitiveness*. Washington DC, USA, Island Press.

Sassen S. (1991) *The Global City*. Princeton, NJ, Princeton University Press.

Sassen S. (2001, 2nd edition) *The Global City*. Princeton, NJ, Princeton University Press.

Scott A.J. (ed.) (2001) *Global City-Regions: Trends, Theory, Policy*. Oxford, Oxford University Press.

Self P. (1967) Regional Planning in Britain: Analysis and Evaluation. *Regional Studies*, 1, 3–10.

South East Economic Planning Council (1967) *A Strategy for the South East: A First Report by the South East Economic Planning Council*. London, HMSO.

South East Joint Planning Team (1970) *Strategic Plan for the South East: A Framework Report by the South East Joint Planning Team*. London, HMSO.

Taylor P.J. and Pain K. (2007) Polycentric Mega-city regions: Exploratory Research from Western Europe. In *The Healdsburg Research Seminar on Megaregions*, Todorovich, P. (ed.) The Healdsburg Research Seminar on Megaregions, Healdsburg, USA, Lincoln Institute of Land Policy and Regional Plan Association, 59–67.

Tewdwr-Jones M., Phelps N. and Freestone R. (2014) *The Planning Imagination: Peter Hall and the Study of Urban and Regional Planning*. Routledge.

Werff M. van der, Lambregts B., Kapoen L. and Kloosterman R. (2005) *POLYNET Action 1.1: Commuting & the Definition of Functional Urban Regions: The Randstad.* London, Institute of Community Studies/The Young Foundation & Polynet Partners.

Chapter 6
Creative Destruction, Long Waves and the Age of the Smart City

Michael Batty

6.1 Introduction

One of Peter Hall's main themes in his research was the impact of technology on cities and regions. Although his early work was largely about the form and function of cities, particularly world cities, and how the planning system in Britain and America was changing the shape of cities, his first visits to the Far East energised his interest in the way cities were crucibles of creativity and innovation. From this, his interest in how long waves of technological innovation since the onset of the industrial revolution have impacted on cities led to an exploration of the geography of Kondratieff cycles and with this, the way in which new technologies destroy the old in the manner first articulated by Joseph Schumpeter. This essay reviews these ideas and Peter's contribution to them but also speculates that the Sixth Kondratieff wave which has just begun will be a culmination of the previous ones in that new information technologies will lead to massively decentralised devices implanted in ourselves and in our cities. In terms of our interest here, the Sixth Kondratieff will be the Age of the Smart City, something that Peter Hall was directing us towards.

Peter, to my knowledge, never wrote about smart cities per se although he was clearly aware of the rise of the movement that had become quite distinct and popular by the time he passed on in 2014. But what he did pursue as a major theme in his research throughout his academic life was the impact of technological change on cities that in many senses was one of his major preoccupations in charting the evolution of cities and the changing face of planning from his first writings in the 1960s. In his autobiographical essay (Hall 2015), he acknowledges several origins of this interest. From his schoolboy days, the library at his secondary school in Blackpool, England, contained Schumpeter's (1939) massive work on *Business Cycles* and his appreciation of the works of the early 20th century economic theorists

Michael Batty, Professor, University College London, UK, Bartlett Professor of Planning.
m.batty@ucl.ac.uk

R.D. Knowles and C. Rozenblat (eds.), *Sir Peter Hall: Pioneer in Regional Planning, Transport and Urban Geography*, SpringerBriefs on Pioneers in Science and Practice 52, DOI 10.1007/978-3-319-28056-1_6

such as Keynes gave him a deep understanding of the importance of innovation and technological change on contemporary as well as past societies. These ideas remained, however, quite dormant in his research and writings in the 1960s and early 1970s but they positively exploded on his first visit to the Far East—Hong Kong—in the mid 1970s.

He was amazed by the dynamism of these new Tiger economies that had suddenly sprouted up, overnight almost, and were in clear evidence by the late 1960s. No one had ever thought that British colonies such as Hong Kong and the newly independent Singapore would become incubators for the world's high tech, financial services and automated manufacturing that they subsequently have become and no one ever thought that the new global economy would be fashioned, not around America, but around China. Japan was 'number one' in those days (Vogel 1979) but this was predicated on their attention to detail, and their continual scrutiny of quality at every level of their products. Japan was still a closed society in many ways and their specialisms in manufacturing were very focussed, with many areas of their economy rather backwards in terms of automated practices and business ethics.

Perhaps the best way to communicate Peter Hall's surprise at what he saw in Hong Kong is to quote him from his essay (Hall 2015) where he says: "I was totally mesmerised by the energy, the power, the achievement of the place. Most surprising for me was the achievement: expecting a third world sort of country, I found a city that in many ways was more advanced than London. And Singapore, to which I paid a flying visit, was even more impressive: it was a kind of 1960s planners' dream, a British city that never was." He continues a little later in that essay to say: "Reflecting for days and weeks after that first trip, I came to an insight that later I found reflected in the writings of Schumpeter: that the reason these societies were so successful was that they gave scope for enterprise, that is for innovation, …".

To an extent, Peter Hall's contribution to this debate is clustered around his writings about high tech and cities, particularly in Europe and North America. He did not write or research much about the Far East for one of his long term quests was to make sense of the changing economic geography of cities, particularly in Europe and the west in general. But the series of books that he wrote between 1980 and the mid-1990s reflected this interest in growth, technology, and communications that were based on ideas about innovation and creative pursuits in cities. His contributions followed three subthemes in this wider domain: books relating to growth poles which in some senses were the drivers of urban change such as *Growth Centres in the European Urban System* (Hall/Hay 1980), and those relating to high tech concentrations, science parks and science cities such as *High Tech America: The What, How, Where and Why of the Sunrise Industries* (Markusen et al. 1986) and the comparator volume in the UK called *Western Sunrise: Genesis and Growth of Britain's High Tech Corridor* (Hall et al. 1987), edited books in the series on urban form and technological change such as *Cities of the 21st Century: New Technologies and Spatial Systems* (Brotchie et al. 1992), and last but by no means least his writings on the wider theory of technological change relating to how cities evolve as reflected in his book *The Carrier Wave: New Information*

Technology and the Geography of Innovation, 1846–2003 (Hall/Preston 1988). In fact a late entrant to this grouping which did include the Far East, was his book *Technopoles of the World: The Making of 21st-Century Industrial Complexes* (Castells/Hall 1994) which went well beyond science parks, written with Manual Castells at the end of their Berkeley days. We will reflect on these writings in our appreciation which follows but before we do so, let us outline the theories associated with Schumpeter which so fascinated Hall and which drove his interest in this area particularly in the 1980s but more generally throughout his intellectual lifetime, studying and speculating on the future of cities.

6.2 Long Waves in the Urban Economy

The idea that history is cyclical betrays a highly deterministic view of evolution that is not much in accord with the empirical evidence, apart from the fact that cycles do exist although perhaps not in so simplistic a form. Nested economic cycles with different periods and amplitudes can be clearly discerned and can be measured fairly unambiguously using various economic indices but whether or not they form any long-term pattern is arguable. However business cycles can be disaggregated and focussed on cities and regions as much as they can on national and international economies. Building booms and busts, for example, which characterise the growth of all cities, provide excellent examples of one of these cycles and the question that Hall asked himself was whether or not the structure of cities and the growth of the an urban economic milieu was reflected in such cycles, particularly those which related to technology. It was Schumpeter (1939), however, who had picked up on long cycles that reflected changes in technology. He popularised the work of Nikolai Kondratieff, a Soviet economist who had argued in a series of publications in the early 20th century which were collected in his book *The Major Economic Cycles* (Kondratieff 1925, 1984), that there appeared to be long waves or super cycles of approximately 50 years in length which mirrored the rise and fall of new technologies, with the particular notion that one cycle led to another; in other words, that once a cycle had worked its way through from its inception to application (in the case of major technologies), another would begin on the tail end of the previous one. The Politburo however did not like this notion because it suggested that capitalism would always adapt and transform itself away from any long term collapse, Consequently Kondratieff who was quite high profile in the Soviet government service, was quickly dispatched to the Gulag where he died in 1938 but not before his work had been translated and picked up in the west, first by Schumpeter.

Kondratieff's and Schumpeter's discussion of long waves was put into perspective by Kuznets (1953) who massaged these ideas into a fourfold temporal structure. Each wave that began when the technology entered the market place, was a period of intense *innovation* and growth. The wave then entered a second phase called *recession* (but not our current usage of the term, more like a drawing-in of the previous boom); this was actually a kind of consolidation of the impact of the technology. This then turned through stagflation into *depression*, which finally bottomed out into a fourth phase of *revival* or recovery. Various terminologies have been used to define this four-fold structure and the terminology remains somewhat obscure. But essentially these waves were not only technological but primarily of wider import in that they describe the economy as passing through inflation and growth, stagflation, deflation-depression, and then improvement only for the cycle to begin again. In fact Mensch (1975, 1979) mapped the diffusion of technologies that follow the usual logistic curve in the way they grow and penetrate the market onto these long waves, thus providing a somewhat richer and more complete portrayal of these theories of innovation.

There are many other cycles that differ from these long waves as well as from the standard business cycle. Rostow (1960) in his book *The Stages of Economic Growth* proposed that shorter waves were subsumed within a series of much longer waves that mirrored the way industrial society emerged through traditional pre-modern, then an era where conditions for technological take-off were set, then leading to the take-off itself which leads to mass consumption (and production). In fact Rostow's model is now quite dated although the notion of these stages occurring as much longer waves is attractive and possible relevant to the development of post-industrial society. Inside these stages can fit Kondratieff cycles while other forms of urban evolution in terms of the morphological structure of cities and regions such as that due to Vance (1990) are also consistent with the waves.

A casual interpretation of Kondratieff waves (K-waves), which is also one that is widely held, is that we are now at the end of the Fifth Kondratieff. With a period of about 50 years, this interpretation suggests the first one began around 1800, perhaps a little before and encapsulated the age of steam with cotton manufacture being the major technological improvement. The second wave began between 1830 and 1850 lasting to 1880, which was the age of railways and steel, while the third wave from 1880 to 1940 was the age of electricity and the automobile. The fourth Kondratieff, which started at the end of World War 2, was the age of information technology and the computer while the fifth wave which in started with the invention of the PC (personal computer) and networking in the late 1970s is the age of the internet and widespread application of information technologies about the person. The sixth wave which some argue is about to begin, might be termed the age of the smart city although I have never heard anyone describe it as such. Undoubtedly the technologies involved are those of the computer and communications but in many senses, this is a new wireless age where computers are able to communicate with each other through any one at any time in any place. What we call it, of course, will only make sense when we see this period of social history in hindsight.

Peter Hall himself first wrote about the Fifth Kondratieff in his frequently published commentaries in the weekly magazine *New Society* in 1981 and 1983. In fact you have to remember that the transistor invented in 1948 at Bell Labs, the integrated circuit in 1959 at Intel and Texas Instruments and the microprocessor or computer on a chip in 1971 at Intel, had only just led to the development of the personal computer and it was very unclear at the time as to the extent to which the computer as a universal machine would come to dominate the world as it has done. Although Peter Hall like several before him defined the Fifth Kondratieff in terms of information technologies, the sheer scale of what has happened in the last 40 or more years was barely anticipated, although by then he was living in San Francisco and must have known of the Home Brew Computer Club and all else that was happening in the Valley. His two articles in the magazine focussed our collective interest not on Kondratieff waves per se but on their geography. In his autobiographical essay, Peter Hall says of his emerging and burning interest in innovation in cities and Schumpeter's contribution:

> The result was an article in *New Society,* The Geography of the Fifth Kondratieff Cycle published in 1981 and a short piece on the 100th anniversary of Schumpeter's birth, which came out in 1983. I suppose that just as parents secretly have favourite children, so authors have favourite works: these two articles, in particular the second, are my own personal favourites. I think that in the three pages that Paul Barker[1] allowed me, I said a lot about what anyone needs to know about the career and work of this extraordinarily exotic and brilliant man, … (Hall 2015: 17)

Every interpretation of Kondratieff long wave theory suggests a slightly different timing and focus of each of the five waves defined since the industrial revolution began. Indeed there are those who consider that these waves go back before the turn of the 19th century, indeed back to the Renaissance and medieval times. Indeed one suspects that one might be able to find evidence of such cycles in the classical era, in China and elsewhere although any data disappears into the mists of time. A recent and much more up to date interpretation of the Sixth Wave has been developed by Naumer et al. (2010). To do this, they define the previous five waves more generically as temporal intervals from 1780–1830 as the era of machine manufacture of clothing, 1830–1880 as the era of mass transportation, 1880–1930 as mass production, 1930–1970 as the time of individual mobility, and from 1970 to 2010 as the era of information and communications technology. Already the waves are out of synch with previous interpretations, such is the ambiguity of this kind of chartism. Kondratieff's waves are often shown as an idealised time series of related waves and Naumer et al. (2010) produce a rather a good and evocative diagram of our current understanding that we reproduce as Fig. 6.1.

The sixth period, which we have just entered, is focused on miniaturization of information and biochemical technologies, which are being used in social and medical applications in the context of institutional change, which will deal with aging, climate change and newly polarized economic structures. In some senses, I would depart from this by suggesting that the internet is increasingly dominant in this sixth wave where devices are disseminated and implanted in virtual everything that we have a concern about including ourselves and that the era of medical advance will really come into its own in the seventh Kondratieff which will probably occur somewhat faster than those in the past. In this sense, the periodicity of the cycles or waves is shortening and eventually they may collapse into each

[1]The Editor of *New Society.*

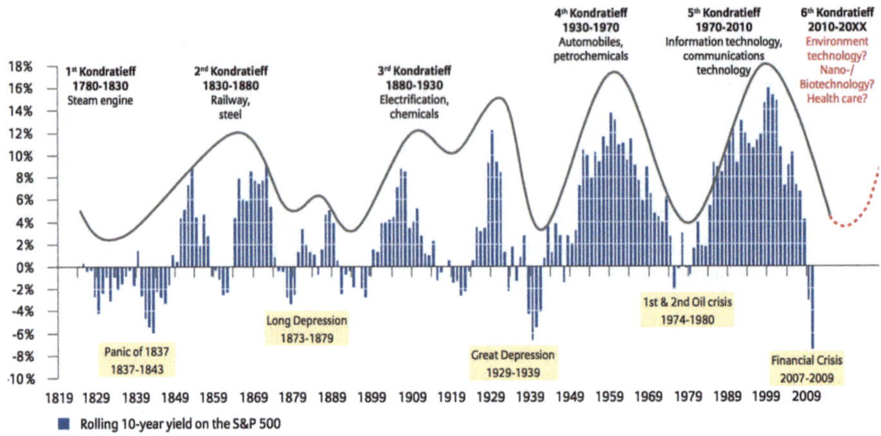

Fig. 6.1 An Interpretation of Kondratieff Waves from Naumer et al. (2010). *Source* Datastream and Allianz Global Investors Capital Market Analysis 2010

other, creating a singularity that is frankly beyond our understanding (Kurzweil 2005). But this is an area we will avoid as it remains in the realms of speculation, no matter however fascinating. Peter never wrote about this and I do not know if he was aware of its import in terms of futurologies, yet it is closely related to ideas about long waves.

There are several features of Kondratieff waves but four main characteristics stand out. Once a technology is developed, it is exploited remorselessly and eventually it peaks, ultimately dying of exhaustion or at least being incorporated into the conventional wisdom and practice of the economy of that time. Human ingenuity as it is however usually advances by beginning to exploit a new technology which is often contained in the genes of the old—just as new scientific theories and paradigms emerge from the limits of that which is pre-existing. The cycle is often dominated in its early stages by an excess of finance capital which again tends to peak as the technology matures usually leading to the third feature which is a severe recession/depression[2] which precedes the upswing in which a new technology is invented and established. During this period, considerable social and institutional transformations can take place as new technologies usually require new organisational structures.

In terms of the fifth and latest Kondratieff for which we might have a complete record and which many agree began around 1970 with the invention of the microprocessor and ended with the invention of the internet, the rapid growth and dissemination of personal computers and networking has dominated the upswing which has also led to the development of the internet in the later growth period. This was widely established in the late 1990s and this led first to the dot.com boom

[2]Enormous confusion reigns about these two words especially in Long Wave Theory. Essentially my own interpretation is that they are loose equivalents of one another.

and bust, equivalent to the period of recession (in the old language of the Kondratieff) and stagflation. In fact this was followed by the great recession which, in Kondratieff terms, is depression from which the world is slowly emerging on a new upswing with very different new technologies in personal lifestyles such as smart phones and apps and early innovations in medical technologies, AI, self-driving cars and so on. This is the food for the sixth Kondratieff that I argue here is the smart city. Or at least some part of this wave is composed of the smart city which is the development of large-scale, all pervasive and invasive computing in personal, public, and private domains and spaces.

Before we conclude this section on the approach to innovation and creativity in technologies that are reflected in cities, we need to examine in a little more detail the way we create, transform and then destroy technologies that are most clearly reflected in the built environment. Schumpeter (1939) argued that not only do long waves start with the creation of new technologies, the very act of creation contains within it the seeds of destruction of the technologies that are replaced or displaced or both. Some technologies, of course, simply become part of the background and are absorbed quickly and often painlessly although their value tends to fall. New technologies sometimes compete directly with old and the process of transformation from old to new depends on strong and fierce, often bitter competition that results in the old being destroyed often with much value within them remaining. Asset stripping is the popular phrase used in mergers and acquisitions. This is particularly true in terms of capital from whichever more value might be extracted even though it contains much intrinsic value in its current form. The best examples tend to be in areas where old technologies become automated and the amount of labour drops dramatically, as processes are computerised: iron and steel, car manufacture and such like are classic examples. These industries do not disappear but simply become leaner, smaller and more competitive, and provide space in the labour market for new technologies to thrive. So the process of destruction is always paralleled or coordinated with processes of creation.

In cities in their built environments, this is even clearer. For example the city of London—the financial quarter—has been rebuilt at least three times since the second world war in times of boom shorter than the Kondratieff for much of this has happened within the fifth wave. It is possible to see perfectly serviceable buildings being pulled down and replaced by ones which embody higher tech but also exploit densities more intensively and thus add value to the stock, reflecting the rapid changes in land values as population continues to rise and place pressure on these most accessible and attractive locations (Batty 2007). In some senses, these changes reflect changing technologies and can be locked onto the long waves that reflect technological change at a macro level. But there are many elaborations of technology that take place on much more rapid cycles and we will note these a little later for they do cast some doubt on the length of the most significant waves in the economy. However it is in the genesis of creation and destruction that lies at the heart of these long waves. Schumpeter himself drew his inspiration from Marx who argued that capitalism is so essentially competitive that new entrants wage war on those who control the current means of production and their competitive instinct is to destroy the old in the creation of the new. In short, those with new ideas always seek to displace and destroy the conventional or received wisdom. This in fact is a truism across society in the construction of knowledge as well as human institutions. It was Max Plank who said that a new generation of science is born only when the old dies out for the old will never accept the new.[3] Old paradigms remain until those who identify with them move on.

There is one last qualification we must make about long waves. There is considerable ambiguity about the beginning and end of each of the waves in modern industrial society, which began in the late 18th century. Technologies can be easily identified in their innovation and development, but how these map onto the business cycle is harder to determine. Schumpeter (1939), like many after him, had a go at this, defining three or more cycles which nested into each other with different periodicities but when we come to look at new information technologies the picture is much less clear. Since the inception of miniaturisation of electronic circuitry that began with the discovery of the transistor in 1948, the rate at which memory and speed has increased has followed the remorseless curve called Moore's (1965) Law.[4] Every 18 months or so, memory and speed double while the cost of such fabrication reduces by half and this has gone on since 1948, which is now somewhat longer than the fifth Kondratieff. Despite the dot.com boom and the great recession, all of which are consistent with the fifth Kondratieff cycle, the power of IT has increased regardless and it shows no sign of stopping. It could be argued too that the emergence of the web and now of highly decentralised devices and apps which are intrinsic to the idea of the smart city, defy the Kondratieff and are not affected by the cycle of creation and destruction which is more of a continuous force rather than a cyclic wave. We will follow these ideas more explicitly below but first we will return to Peter Hall's application of the Kondratieff to geography and to cities.

6.3 The Creative City: Crucibles of Innovation and Incubation

Creativity and innovation go hand in hand in contemporary societies and in the last 30 years, cities have begun to be defined by their ability to foster such creativity and innovation. One of the greatest proponents of this view was Jacobs (1961) who in her seminal work—which was superficially a denunciation of the top-down US planning system which she argued was destroying the American city—explained

[3]Strictly he said: "A new scientific truth does not triumph by convincing its opponents and making them see the light, but rather its opponents eventually die, and a new generation grows up that is familiar with it" (Planck 1950, *Scientific Autobiography and Other Papers*, London, Williams and Norgate, p. 33).

[4]See at: http://spectrum.ieee.org/semiconductors/processors/the-multiple-lives-of-moores-law.

that cities brought together and focussed incredible diversity of talent. It was this variety that led to much greater innovation and creativity than could be assumed to take place in lower density, more rural and possibly poorer agrarian settings, and thus she argued that the quality of life was considerably richer in cities than outside of them. A modern rendition of Jacobs' argument has recently been presented by Glaeser (2012) who takes the argument further implying that the larger the city, the greater the diversity and the better the quality of life for its inhabitants.

Like all polemics, there is a grain of truth in this but there is also considerable disquiet in its argument. Bettencourt et al. (2007) drawing inspiration from Marshall and his definition of agglomeration economies, present evidence that the bigger the city, the greater the *per capita* wealth but so too is the greater the inequality within their urban populations and it is entirely possible that the extreme wealth associated with a very small number of persons dominates the much lesser wealth associated with a much larger number of poor people. In short, the picture concerning big cities and their agglomeration economies and diseconomies is quite confused but what does seem to be clear is that the number of creative activities grows more than proportionately as cities get bigger, especially in employment which can be measured in terms of their productivity by patents, in entertainment and highly creative pursuits in the arts, indeed in any occupation or activity that depends on highly creative talent.

Creativity is an intensely personal act and if cities represent the crucibles of invention and innovation, they are essentially mechanisms for bringing people together from which sparks will fly. It is density and connectivity then that raises the stakes in cities and it was this that Peter Hall first saw in Hong Kong in 1975. Cities are built from the bottom up and in the last 30 years, this has been widely articulated as part of complexity theory (Batty 2005). In fact economic cycles are clearly determined from bottom up action as there is no central control guiding the economy to produce such waves. Such waves are the product of an invisible hand—the product of individual creative and competitive actions that generate optimism and pessimism in the economy, and the fact that there are upswings and downswings is entirely consistent with the notion that new inventions are spurred on by positive feedback between those working on them while when the technology matures, the market adapts to their presence, assimilates them and their early promise becomes widely accepted with the initial enthusiasm and added value wearing off. This sets off recession and thence depression as things overshoot and the early promise disappears. This of course drives individual inventors and entrepreneurs to try to renew their technological edge and the search for new technologies and new markets starts the cycle over again. Positive feedbacks kick in at every stage.

Peter Hall's most significant book on the Kondratieff written in 1988 with Paschal Preston is called *The Carrier Wave* (Hall/Preston 1988) and is really a history of 'New Information Technology and the Geography of Innovation, 1846–2003' which is the subtitle of this work. In fact, his book is not really focussed on cities and regions but more on the forces of information technology from the telegraph to the personal computer that have driven the third, fourth and fifth Kondratieffs. In one sense, Peter needed to write this book before his last significant text on these new technologies, namely his *Technopoles of the World* (Castells/Hall 1994), which does in fact focus on the geography of where such technology is produced or rather invented. What Peter's work on the Kondratieff really shows is something that others have not really focussed upon: the waves can be quite uneven geographically and this of course blurs their temporal coincidence. Hall and Preston argue that by the end of the third Kondratieff, Britain as an industrial nation was lagging in comparison to its economic counterparts. This is a hard argument to demonstrate unequivocally for the radio, TV, and then the digital computer were much in evidence in terms of British inventions during the fourth Kondratieff. Yet during the fifth, the British economy lost whatever advantage it had in new information technologies. What other nation could invent one of the world's first digital computers *The Colossus* to crack the German U-boat codes only for the ten that were on order for Bletchley Park to be broken up for scrap at the end of the Second World War before they had been delivered to the code breakers. Turing's message that such machines were universal—that they could be used to do anything that could be reduced to binary digital code and most things can—has taken a long time to take sink in. This more than anything else defined the fact that the fifth Kondratieff would barely happen in Britain as its economy headed downwards to de-industrialisation although the sixth wave shows every sign of restoring the balance.

A geography of innovation of a kind was produced by Castells/Hall (1994) that in some respects is the obvious complement to *The Carrier Wave*. This is not really a geography of how new information technology is being used as it develops during the fourth and fifth Kondratieffs but a geography of the places where new information technologies are produced—or rather invented and then produced which are often in different locations. They argue that the focal points where new IT has been invented have been in planned rather than spontaneously evolving locations but I doubt this very much and find it a strange argument. In fact, where these new technologies are produced is not in the core or inner areas of cities where earlier technologies were located. Indeed new industrial zones have been created, the classic example being in Silicon Valley, which has evolved from the bottom up, a product of massive path dependence due to a succession of historical accidents accompanied by positive feedbacks that have built many of the world's most prominent computer companies. Some nations have identified science parks and subsidised these while others have identified entire science cities but once begun perhaps with some seed funding a little like new towns and garden cities before, these initiatives have become self-sustaining. In fact what constitutes planned and what constitutes organically evolving spatial structures can be quite blurred as most development is a combination of both, neither entirely bottom up or top down.

There is nothing very specific about the locations where the most innovative ideas are produced. In fact the most basic of ideas tend to be produced almost randomly in space and time. Clusters of innovations do tend to occur in high density situations in and around the core of large cities but these tend to cluster around universities too which tend to be quite central in their location. Innovative ideas are also generated in suburban locations, in the sprawl as well as in these dense clusters and regionally such locations tend to depend more on the general economic milieu than on specific fine scale locations per se. So for example most new IT has been produced in the south of Britain or in its biggest cities and universities, and the same is true in the United States. Insofar as these patterns are not borne out, then it is historical accidents that are the determining factors.

Peter Hall never really discussed how we use information technologies in working and living in cities and to an extent the smart cities movement is much more concerned with these kinds of issue. It matters little where such technologies are produced but how they are used and how they change our behaviour and in terms of the way cities are organised and develop, this is all-important. In Peter's quest to explain technologies and their role in structuring and enhancing the quality of life in cities, his focus was mainly on how such technologies inform and enhance the economic quality of life in cities. In fact, technologies also divide the population and some of the wider issues concerned with populations which are segregated into information rich or poor, relate to how cities are structured spatially. Peter's work is mainly on the supply of technology, on its production not on the demand for technology and they way it is used.

What I will do in the rest of this chapter is speculate a little about how I think Peter Hall might have reacted to the smart cities movement had he lived longer. I think that the movement is so profound and deep-seated, in many senses so misguided and so all embracing, that he would have begun to react and comment on its importance and impact. In one sense, he had already begun to embrace its effects in his focus on communications technologies. Reades (2016) in this volume explored many of these questions in his doctoral thesis and he was one of Peter's last students who stood astride the divide between substantive issues of technology in cities and the way those same technologies are used by ourselves to understand cities. In his chapter here, Reades articulates Peter's key approach to cities and technology when he says: "Peter did not, I think, have a lot of time for purely econometric reasoning; it was relevant principally insomuch as businesspeople and planners thought it was when taking decisions" (see Hall 1962: 169). It was, in other words, generally better to look to history for both the bigger trends and recurring patterns, as well as for the details. So history as recounted here in terms of the Kondratieff is the key and the more pragmatic usage of technology with respect to the way it alters human behaviour which is what big data and the smart city is all about is something that Peter would only have commented on when much more evidence of its meaning had become available. Just as he commented on the British planning system in his volumes on *The Containment of Urban England* (Hall et al. 1973) from a historical perspective, he would have begun the same on the smart cities movement in the near future. The fact that he did not live long enough to be able to do this, gives me a chance to make my interpretations of what he might have said with respect to the current development and applications of this technology which define what many speculate is the sixth Kondratieff.

6.4 The Sixth Kondratieff: The Age of the Smart City

In fact, Peter did talk a little about smart cities in a more popular way, although this can hardly be called research. He was well aware that computers were continually entering all facets of daily life and in this sense automating the city beyond anything that had been speculated upon in the 20th century—beyond Fordism, beyond automation (which is now a somewhat dated term pertaining to the industrial era), beyond mass production, even beyond niche manufacturing. He saw this particularly in new forms of behaviour pertaining to transport which in some senses is the heartland of the smart cities movement. In his popular journalism, he referred extensively to such technologies talking about how families of the near future will juggle travel and job locations using smart phones which are programmed to keep them on track, literally as well as metaphorically (Hall 2010). To an extent what Hall and others perceived is that these kinds of technology are really part of a seamless array of technologies that we as planners and geographers need to consider in a more integrated way than we have in the past: thinking not only about the production and consumption of technologies but how we use them to understand that same production and consumption.

It is worth emphasising this point. Peter's work on technology was very much in the spirit of economic and industrial geography, the geography of why, where and how different technologies were located with respect to their production. In this sense his work is in the tradition of location theory, which tended to deal with production and supply rather than consumption and demand. This latter focus on consumption and demand is more evident in analysis of the information economy with respect to the way we as individuals are using new technologies. As in any study of the geography and structure of the economy, there is an asymmetry between the way we study production and consumption, demand and supply with these two foci being poles of study that define our knowledge of how new technologies are invented, disseminated and consumed.

In terms of cities, there are two other dimensions that pertain to technology. In fact new information technologies first made their appearance not in the study of automation in cities or in the geography of automated products but in the use of computers for planning cities. In fact, my own work is very much in this domain and only touches on the notion of the geography of production insofar as our models reach out to include such geography in the models that we are using to simulate such form and function. While this dimension has been progressing with ever more sophisticated, ever faster models, with better and better visualisations, computer technologies are now finally being embedded into the fabric of cities. Computers began in the scientific lab, moved quickly to organisations in the 1950s for transactions processing, hit the personal and entertainment market in the 1980s, while all the time being networked in terms of remote processing and storage. Now computers have begun to be embedded into the public domain and that is where we see the advances in transportation technologies and the demand for them so cogently described by Peter Hall in 2010 in his thinking about the near future.

This is essentially our third domain with respect to geography and planning—the smart city which essentially is this embedding of computation and communications into the city so that it might be made more efficient and possibly more equitable although this is hardly guaranteed. In terms of its definition and scope, the smart city also includes questions about urban data—big data—and it extends to analytics, which at a push might include urban models of the kind that pertain to forecasting and impact analysis. The key dilemma that my own field is facing is the extent to which time frames are changing in terms of our understanding of the city due to a focus in the smart city on more routine, real time change—not a bad thing and long overdue—but something that is taxing our abilities to explain and simulate. There are also strange inversions about how we can use our technologies to study the same technologies we are trying to explain and thus in how they are changing the system of interest, the city. This new technological milieu in cities, their planning and their geography is now the focus.

The main feature that is changing our world and the world of cities that Peter wrote about is the fact that cities are becoming more complex as they evolve. In short, the subject of our interest is forever shifting and this means we have to run to stand still. This of course is social change but there is little doubt that material wealth as well as new communications technologies is massively broadening our personal and collective horizons and this is making cities harder to understand, more heterogeneous, with more complicated order and pattern as well as a multiplicity of new ways of production and consumption. The smart city is just one manifestation of this increasing complexity. What new information technologies are doing is that when embedded into the public domain—in transport, energy, and related utilities as well as the ways we access them and control them—they are generating massive amounts of data about the functioning of the city which is the 'exhaust' of real time processing. The big data that comes as part of this exhaust is highly unstructured, some of it is noise but it is giving us a temporal perspective on the city that we have never had before. Peter did not have the opportunity to rely for any of his research on the big data that comes from such real time technologies but he was aware of what was possible and what he himself needed to make sense of the future form of cities.

To a large extent, Peter's focus on future cities was related to the creative cities, to cities which were being built around new media drawing on the work of a variety of ideas from those associated with Florida (2002) to Landry (2008). In fact since these ideas began, the entire world of cities has been revolutionised by social media mainly associated with smart phones where access to all of use (who have such devices) has produced new layers of complexity in the form of social networks, either directly as through media such as *Twitter* or through web site such as *Facebook*. Peter did not write anything about this kind of media as far as I know for we are all still getting to grips with its meaning with respect to how cities form and function. Vast new data sets are available in real time that tell us where and what we are engaging with using this media but our understanding of their meaning is extremely primitive. There are many open questions as to what such media means with respect to the organisation and planning of cities and we stand at a threshold with respect to how the future city will embrace such new technologies. It is this as much as anything else that marks the beginning of a new Kondratieff, the Sixth, which we have penned as The Age of the Smart City.

In the lecture he gave to the Balzan Foundation (Hall 2005) where he received the prize for his book *Cities in Civilization* (Hall 1998), Peter Hall ends by saying that what he wants to do is to supplement and extend his work on urban polycentricity called POLYNET that had occupied him in his last decade (Hall/Pain 2006). He wished to compile and interpret much better and more complete data on information flows. He says: "There is a parallel research agenda, impelled by the one piece of the POLYNET research that proved a relative failure. This was the attempt directly to measure the flows of information between firms and between places in such polycentric regions, by measuring both business travel and telecommunication flows—telephone calls, email messages …. Since completion of the study, however, I have discovered a rich potential source: the geography of mobile telephone traffic" (Hall 2005). It was this that Reades (2016, this volume) worked on for his doctorate and indeed this propelled Peter towards smart city ideas for Jon worked with ourselves in CASA on these issues and now has taken up a part of Peter's research agenda. I think that in a way this might have drawn my own group back towards Peter as it was slowly doing and it is to my eternal regret that our busy lives in the last few years did not provide us with the opportunity to talk about the smart city, about social media, about new forms of network and networking and the future of the city more generally. All I can do in concluding this essay on Peter's contribution to the geography of technology and innovation is speculate on what might have been.

We did in fact write (only) one paper together (Batty et al. 1974) when the idea of fares-free public transport was much in evidence and we were working at Reading University. I cannot remember the wider context but it involved the West Midlands where this idea was very much in the air. It was David Starkie who brought us together. I developed the land use transport model which essentially reduced transport costs to zero for a simple public-private two mode model of Birmingham and we engaged in some predictive modelling, not a million miles away from what we are still doing with much more detailed data pertaining to fares on public transport systems in London using RFID data, big data from the Oyster card smart card data set that we have. Our 1974 paper was hardly in the tradition of the smart city for in those days even the idea of the PC was unknown and this was still the era of interacting with computers using a deck of punched cards. But if you look back to what we did, then you will find some ideas that relate technology to model building to new policies about public transport that anticipate the smart city and its analytics (Batty et al. 1974). Clearly Peter had thought hard and long about the implications of all this for many years.

One of the key features of the role of technology on and in cities involves the ever changing influence of distance and its translation into the costs incurred to moving in time and space. Once the fifth Kondratieff really hit the modern world by the 1990s with the widespread dissemination of the PC and the development of the internet, the prospect of the 'death of distance' as popularly articulated by

Cairncross (1997) and the rebirth of Toffler's (1970) electronic cottage where everyone worked from home began to tie all these approaches to the new technologies together. What has not happened is the explosion of the city into far-flung bits but the global networks that now tie everyone together in a myriad of ways have made the world infinitely more complex as we implied above. It is this that complicates the notion that the smart city will become a seamless set of interfaces between a multitude of information technologies. In fact we can no longer discuss the production of information technologies without recourse to examining global locations, nor can we explore the behaviour of people in large cities without looking at their global connections and the costs of their interacting with others at a distance. Physical travel is as costly as ever relative to income and job type but ethereal travel—which is a shorthand for the way we interact through information—email, Skype, the web, all kinds of social media/networking—confuses the geography of the future city dramatically. In fact to study cities in the future, we will need every one of these different approaches to technology that we have identified in this section. This is something that Peter Hall would have signed up to, of that there is little doubt, and this what makes his contributions through his many books and articles of enduring value.

6.5 The Technological Future

Several commentators have argued that the sixth Kondratieff is likely to be marked by rapid technological advances in the biosciences—in medical technologies that will be key to expanding our life limits, and curing long standing diseases. But this will be built on the back of new information technologies developed during the fifth and in this sense one might think of the smart city as being just another way of characterising this change. Biotechnologies and nanotechnologies will doubtless be key to these developments and so will social media but in principle, these developments have the power to change the entire nature of the Kondratieff itself. As we have implied the periodicity of the typical cycle itself may well be shortening but in one sense, the wave is so long that it embraces several shorter cycles and in one sense, we might simply abandon the notion of the long waves and focus on merged continuous shorter ones. But at the end of the day, there can be no disputing that fact that the industrial revolution itself divides into early and late and the post-industrial and the Kondratieffs coincide to a degree with these long historical periods.

One feature of this technological future is the changing structure of demand and supply. Much of our past understanding of the geography of technology pursued by Peter Hall was focused on production rather than consumption. Increasingly the technologies of the future will be individual rather than collective; health is clearly a case in point. Already education has increased dramatically in modern cities and now the geography of education is significant—this still tends to be concentrated but increasingly education and health will be decentralised and this has profound implications for the smart city. To an extent, the geography of future cities which Peter Hall was so much involved in will be very different, much more decentralised and heterogeneous than the cities of the industrial past. Indeed it is worth concluding with his opening remarks in his address for his Balzan prize. When talking of the mysterious nature of life, he said: "I've spent my academic life asking "How do cities work?" and I don't think that I've yet cracked my mystery either". That is as maybe but for what he has done, we are all a lot wiser of about a path best followed and we owe him thanks for the directions to the future that he left us with.

References

Batty M. (2005) *Cities and Complexity: Understanding Cities with Cellular Automata, Agent-Based Models, and Fractals.* Cambridge, MA, USA, The MIT Press.

Batty M. (2007) The Creative Destruction of Cities. *Environment and Planning B*, 34, 1–4.

Batty M., Hall P., and Starkie D.N.M. (1974) The Impact of Fares-Free Public Transport upon Urban Land Use and Activity Patterns. *Transportation Research Forum,* 15, 347–353.

Bettencourt L., Bettencourt L.M.A., Lobo J., Helbing D., Kuhnert C. and West G.B. (2007) Growth, Innovation, Scaling, and the Pace of Life in Cities. *Proceedings of the National Academy of Sciences USA,* 104, 7301–7306.

Brotchie J., Hall P., Batty M., and Newton P. (eds.) (1992) *Cities of the 21st Century: New Technologies and Spatial Systems.* Longmans Cheshire, Melbourne, Australia.

Cairncross F. (1997) *The Death of Distance: How the Communications Revolution Will Change Our Lives.* Cambridge, MA, USA, Harvard Business School Press.

Castells M. and Hall P. (1994) *Technopoles of the World: The Making of 21st-Century Industrial Complexes.* London, Routledge.

Florida R. (2002) *The Rise of the Creative Class. And How It's Transforming Work, Leisure and Everyday Life.* New York, Basic Books.

Glaeser E. (2012) *The Triumph of the City: How Our Greatest Invention Makes Us Richer, Smarter, Greener, Healthier, and Happier.* New York, Allen Lane, The Penguin Press.

Hall P. (1962) *The Industries of London since 1861.* London, Hutchinson & Company.

Hall P. (1981) The Geography of the Fifth Kondratieff Cycle. *New Society*, 26 March, 535–537.

Hall P. (1983) The Third Man of Economics. *New Society*, 1 December, 343–344.

Hall P. (1998) *Cities in Civilization.* New York, Pantheon Books.

Hall P. (2005) *Apologia Pro Vita Sua.* Address to the Balzan Foundation, available at http://www.balzan.org/en/prizewinners/sir-peter-hall/apologia-pro-vita-sua.

Hall P. (2010) A Day on the Trains, 2030—the Dumill-Trents. *Town and Country Planning,* 79(1), 8–10.

Hall P. (2015) A History of Contemporary Planning: Personal Stories, Personal Journeys in Britain and America. in Haselberger B. and Saija L. (eds.) *Encounters in Planning Thought: Autobiographical Essays from Key Thinkers in Planning.* forthcoming.

Hall P., Breheny M., Hart D., and McQuaid R. (1987) *Western Sunrise: Genesis and Growth of Britain's High Tech Corridor.* London, Allen and Unwin.

Hall P., Drewett R., Gracey H. and Thomas R. (1973) *Containment of Urban England.* London, Allen and Unwin.

Hall P. and Hay D. (1980) *Growth Centres in the European Urban System.* London, Heinemann.

Hall P. and Pain K. (2006) *The Polycentric Metropolis: Learning from Mega-City Regions in Europe.* London, Earthscan.

Hall P. and Preston P. (1988) *The Carrier Wave: New Information Technology and the Geography of Innovation, 1846–2003.* London, Unwin Hyman.

Jacobs J. (1961) *The Death and Life of Great American Cities*. New York, Random House.
Kondratieff N. (1925, 1984) *The Major Economic Cycles*. Translated by Daniels G. in 1984 as *The Long Wave Cycles*. New York, E.P. Dutton.
Kurzweil R. (2005) *The Singularity is Near*. New York, Viking.
Kuznets S.S. (1953) *Economic Change: Selected Essays in Business Cycles, National Income, and Economic Growth*. New York, W.W. Norton Inc.
Landry C. (2008) *The Creative City: A Toolkit for Urban Innovators*. London, Routledge.
Markusen A., Hall P., and Glasmeier A. (1986) *High Tech America: The What, How, Where and Why of the Sunrise Industries*, Boston, MA, USA, Allen and Unwin.
Mensch G. (1975, 1979) *Stalemate in Technology: Innovations Overcome the Depression*. Cambridge, MA, USA, Ballinger Publishing Company.
Moore G.E. (1965) Cramming More Components Onto Integrated Circuits. *Electronics*, 38, 8, April 19, 114–117.
Naumer H.J., Nacken D., and Scheurer S. (2010) *The Sixth Kondratieff—Long Waves of Prosperity*. Allianz Global Investors, Allianz Global Investors, Kapitalanlagegesellschaft mbH, Mainzer Landstraße, Frankfurt am Main, Germany.
Planck M. (1950) *Scientific Autobiography and Other Papers*. London, Williams and Norgate.
Reades J. (2016) *Location & Innovation*, Chapter 3, this volume.
Rostow W.W. (1960) *The Stages of Economic Growth: A Non-Communist Manifesto*, Cambridge, UK, Cambridge University Press.
Schumpeter J.A. (1939) *Business Cycles*. New York, McGraw-Hill Book Company.
Toffler A. (1970) *Future Shock*. New York, Random House.
Vance J.E. (1990) *The Continuing City: Urban Morphology in Western Civilization*. Baltimore, MD, USA, Johns Hopkins University Press.
Vogel E. (1979) *Japan as Number One: Lessons for America*. Cambridge, MA, USA, Harvard University Press.

Chapter 7
The Visionary of World and European Cities

Céline Rozenblat and Daniel O'Donoghue

7.1 Introduction

Sir Peter Hall was one of the outstanding urban geographers of the latter half of the twentieth century. His name is synonymous with cities and urban planning. One can only marvel at the depth and breadth of his contributions to the understanding of cities that, for 50 years have enlightened our thoughts of cities past, present and future. We never worked directly with Sir Peter Hall, but as custodians of the Commission for Urban Geography of the International Geographical Union (IGU) we owe a huge collective debt to Sir Peter on behalf of our Commission and all urban geographers for the direction his life's work on cities and systems has taken us. As far as we know, Peter Hall never frequented the IGU Urban Commission's meetings, yet he inspired many of the Commission's discussions and orientations. In return, a part of his inspiration derived from some of the major contributors to the Urban Commission during the 1970s and 1980s, for example the Canadian geographers Larry Bourne and Jim Simmons. The 1960s witnessed Brian Berry and Richard Preston, among many others, become leaders of a newly emerging "quantitative urban geography" that would nourish the reflections of both Peter Hall and the Urban Commission's members.

Peter Hall was a pioneer and visionary in the comparative approach he adopted to the study of cities around the world, and in particular, European cities. His approach was quite holistic, linking the dynamism of cities to their economies, their technologies, their governance, their culture and art (civilization) and their well-being. We have left a detailed review of his work to the specialists he worked

Celine Rozenblat, Professor, University of Lausanne, Switzerland, Associate Professor of Geography; Celine.rozenblat@unil.ch.

Dan O'Donoghue, Dr., University of Canterbury in Kent, UK, Principal Lecturer in Geography; daniel.donoghue@canterbury.ac.uk.

R.D. Knowles and C. Rozenblat (eds.), *Sir Peter Hall: Pioneer in Regional Planning, Transport and Urban Geography*, SpringerBriefs on Pioneers in Science and Practice 52, DOI 10.1007/978-3-319-28056-1_7

with and those who knew him best. However, we must appreciate the nature of the common themes and threads that were interwoven throughout his works. Many of his visions became reality and to understand better why his visions of cities were so relevant, we must recognise that:

1. He acknowledged the unique nature and role of place yet continuously searched for universal urban processes that could unify our understanding of all places. He particularly managed to achieve this goal through a comparative approach;
2. Using a variety of regional, national, continental and international samples of cities, he explored the ever increasing linkages between urban transformations and technological change;
3. He assumed urban development was an incremental or staged process and thus recognised the role of history in urban evolution, studying long and short term transformations;
4. He adopted a multi-scale approach to the study of cities contextualizing urban trends giving high importance to the impact of exogenous regional, national and global forces and influences, but also a multi-level approach that consistently linked intra and inter-urban processes.

Sir Peter Hall clearly understood the complexities involved in any attempt to understand what made cities 'tick' but what separated him from most other researchers was his ability to use his knowledge to create policy that could address real world problems. He had the ability to evaluate the possible impacts of urban and economic policy and it could be argued that his ideas have transformed planning and urban landscapes around the world. The remainder of this chapter will explore the four persistent themes, identified above, that underpin his body or work.

7.2 Universal Urban Processes and the Uniqueness of Cities: Comparative Approaches

His first aim, one that remained subjacent to most of his work, was to understand and identify the forces behind metropolitan growth and the political responses produced by local and national governments. To address these central questions, he developed a range of different strategies in his numerous books, however one of these strategies appeared with regular frequency: the comparative approach. When he explored *World Cities* (Hall 1977), with chapters dedicated to each city, his main interest was to reduce the range of forces for all of the cities to some common denominators thus isolating general forces that might be developed into a vision of 'the future metropolis' in the final chapter. Through an analysis of seven metropolises (New York, London, Tokyo, Paris, Moscow, Randstad, Rhine-Ruhr) he aimed to identify forces, even allowing for contextual deformations that would converge and result in some comprehensive general model. This model could be adapted to the irregularities of each city yet allow the underlying driving forces to continue to concentrate people in these metropolises.

The seven metropolises he chose did not form some comprehensive or exhaustive list, these were not the only places that could be considered as World Cities, but they served as a set of diverse examples of urbanity that when studied and compared collectively, permitted a better understanding of the likely future direction development might take in many other cities. He also adopted the comparative approach in Cities in Civilization (Hall 1998) where in different sections (books) he explored: 'the city as a cultural crucible'; 'the city as an innovative milieu'; 'the marriage of art and technology', and; 'the establishment of an urban order'. In these sections he compared various attributes that made these urban places, but he complicated the approach by not just comparing these attributions over time, but by comparing different locations from across the world over different time spans. Thus the great ages of Athens, Florence, London, Vienna, Paris and Berlin are explored in a cultural setting while Manchester, Glasgow, Detroit, Silicon Valley and Tokyo are placed in a technological paradigm. Los Angeles, Memphis, Rome, New York and Stockholm also find themselves examined in other contexts. All the comparisons combined to provide insights into the diverse and often time specific forces that lead to the emergence of each unique place, while also identifying the common qualities, e.g. culture or technology that explains the emergence and expansion of all of these cities. He systemized this method, applying his knowledge of each set of forces across a range of places to ultimately expressing his optimism in the fifth and final section of the book (Hall 1998), entitled The Union of Art, Technology and Organization, where he wrote of 'the Coming of the Golden Age'.

In order to pursue more systematic comparative studies Peter Hall recognised there could be some major problems, particularly with the different nature of cities' functions.

> For every major country of the world it should be possible, with the aid of the original metropolitan area definitions from International Urban Research, to distinguish a distinct metropolitan region: a single dominant metropolitan area, surrounding and growing out from a world city. Sometimes, indeed, in some highly centralised states of Western Europe, this is perfectly simple (…). But difficulties emerge when the administrative capital is separated from the commercial and financial capital… (Hall 1977: 13–15).

Moreover, in order to rigorously compare metropolises, he pointed out the limits to direct comparisons of urban regions if appropriate and consistent data were not available thus leading to potentially misleading conclusions.

He went further with Hay for the study of *Growth Centres in the European Urban System* (1980), giving a special attention to the comparable delineation of cities. Here, the third chapter ('The Technical Building Blocks' [pp. 32–83]) is entirely dedicated to the make up of the Urban Systems Regions, detailing country by country how each census considers the population, how employment and commuting are relevant, and the problems of temporal comparisons (important for longitudinal and for international comparisons). Their work encountered many problems in both defining urban regions and their surrounding areas and when the lack of comparative data put limitations to results, it seems to have frustrated the two authors immensely thus leading to a plea for better cooperation for data at the European scale (p. 233). This is still an issue for urban geographers today, and the comparative approach has become embedded to the extent that agencies around Europe have worked hard to improve this still unsolved urban data problem for modern researchers.

The *Polycentric Metropolis* (Hall/Pain 2006) was another systematic comparative piece of work on European cities, a quarter of a century after the book with Hay in 1980. The comparison of 8 European mega-city regions was also based on a deep and rigorous reflection on the issue of urban delineations. Dissatisfaction with a lack of consistent urban definitions and boundaries was still evident when the authors note in the conclusion,

> Polycentricity depends on scale: the concept of polycentricity is scale dependent and cannot be simply mapped on to fixed Mega City Region configurations, such as the ones used in the study (p. 198).

However, the scaling dependence here is not only a methodological problem. Rather, it introduces a larger conceptual question regarding cities whereby urban functions are uneven across space and place and where their reach, range and influence over both distance and time varies significantly. Ironically, the search for common features is more difficult due to the very individuality of cities and their specific functions that create the size and delineation problems that prevent easy comparison.

Perhaps it is stating the obvious to say that data and definitions for cities, regions and nations needs to be standardised across cities if direct and objective comparisons are to be made between places. In Hall/Tewdwr-Jones (2010) the associated problem of complexity and comparability within and between cities, is dealt with succinctly. With particular reference to planning, and by implication cities, the authors point to the complexity of planning. The recognition that there may be dozens of objectives in any one plan with many of those either not readily comparable or even contradictory, demonstrates the problem of comparative approaches. Given the complexity, and lack of certainty often found in the physical sciences, Hall/Tewdwr-Jones (2010) relate the study of cities and spatial planning to the other social sciences whereby 'we have to work with laws of statistical tendency rather than the laws which are constantly reliable in producing experimental results' (p. 12). By introducing probabilities to the mix one must beware of the quagmire that might emerge as one goes deeper and deeper into the data in search of answers.

Yet, Peter Hall seemed to know intuitively that the more information you could find for more places, the stronger the statistical evidence, and the more likely it would be to identify universal urban processes in the face of the incredible diversity and uniqueness of places. It is the attempt to try to reconcile all the disparate dimensions and objectives across a network of places that is the essence of urban and regional planning. Unless there is recognition that individual places are analysed along certain dimensions *vis a vis* other locations, it is impossible to plan. It is implicit in Hall's work therefore that a comparative approach is a minimum requirement for any attempt to understand the identity, role, function and performance of cities and regions.

7.3 Urban Transformations and Technological Change

Peter Hall identified and considered cities not just by their population, but more importantly by the functions and activities undertaken in those places. These were places where production, consumption, trade and innovation, took place within a wider network of linked places. Cities are places where technological change occurs with a strong link to innovation (Hall 1985). This approach is evident throughout his canon of work but it is more deeply developed in *The spatial impact of technological change* (Brotchie et al. 1987) and in *Urban Future 21* (Hall/Pfeiffer 2000): "Everywhere, cities are great engines of economic growth. Their share of national output, in almost every country, is much higher than their share of labour force. Cities make their people productive" (p. 51). He explains this by the "*system of global competition*" where cities permit accumulation in market economies, combining global and local markets. In this process, the succession of technological waves (see also Batty in this volume) took a large part of Peter Hall's attention (Hall 1985). It is clear at this stage that Hall was well aware that technology itself was changing and its transformative capacity was about to change radically as an informational economy would require more and more informational technologies, with their concomitant impact on social organisation and space.

Peter Hall specified the nature and the order of magnitude of the substitution of manufacturing production by services, insisting on the new forms of activities and organizations and their spatial implications. In this way, he identified earlier than Sassen (1991), the potential effects of the substitution of services organizing production in the big metropolitan centres and their increasing concentration (Hall 1987a). His approach highlighted economic and planning effects of this concentration, whereas Saskia Sassen insisted more on the social and financial aspects. Comparing the geography of High Tech in Great Britain and USA, he interpreted the gap between USA and Great Britain and their technological evolution contextualizing them thus,

> The difference in essence is this: in the United States, high tech contributed 87% of an impressive 1,250,000 gain in manufacturing jobs over the decade; in Britain, high tech actually contributed to a catastrophic loss of 1,963,000 manufacturing jobs—almost exactly one-quarter of the 1971 total. The best that can be said is that high tech helped a little to reduce the overall rate of decline (Hall 1987b: 144).

Trying to explain concentration, he mobilized *agglomeration economies* as a process that may be highly significant for small firms in the early stage of an industry's development. In particular, the concentration is due to the sharing of services, information, or skilled labour.

> This, however, still leaves the critical question: why, apparently, does high-tech industry now reject older central cities, thus denying the seedbed function once thought critical for places? This suggests a central weakness of the traditional theories (Hall 1987b: 151).

Exploring Great Britain and the USA and the role of military and defence establishments after the World War II, he concludes:

> The development of high Tech Industry in the new industrial regions is not a random process. It arises from conscious decisions to locate R&D and related procurement facilities at certain times, often under military stress in time of war or defensive preparation, and with a strong strategic component in the locational decision (Hall 1987b: 153).

Forces shaping Urban Europe (Hall 1993) is another example of his awareness of the integrative and comparative approach to understanding cities especially regarding the role played by technology within an urban system's context. Unlike many authors of the era still obsessed with production, Hall clearly understood very early the transformative role of technology within the service sector of the economy, particularly with reference to the informational economy. Amongst the many forces he identified in that paper, he paid particular attention to the centralizing power of high level services brought about by a shift towards employment in the informational sector.

Working with Castells a year later (Castells/Hall 1994) the role of technology is to the forefront of his thinking. Once again, a comparative approach is in evidence exploring disparate types of places across a range of cities and countries. They explored different High Technology in Silicon Valley, Boston's Highway 128, *Science Cities* in Akademgorodok in Siberia, Taedok in South Korea and Tsukuba and Kansai in Japan, *Technology Parks* in Sophia-Antipolis, France, Cambridge in England and Hsinchu in Taiwan (note the change of scale), *Innovative Milieu* in London, Paris and Tokyo, *Metropoles* in Munich and Southern California, amongst many other places. They then integrated all of the information in an attempt to distil the lessons from all of these places in the implied hope of developing appropriate strategies that might create the successful *Technopoles* of the future. In fact one could argue that along with other well-known researchers like Krugman or Porter, Hall's work was a significant contribution in establishing high technology clusters as the 'Holy Grail' for a generation of urban and regional planners.

7.4 Stages in Urban Development

Despite acknowledging the specificity of historical events and political decisions in the urban dynamic, Peter Hall continuously searched for regularities in Urban Systems development. One of his best demonstrations, is the study of the European Urban system with Hay (1980), where they evaluate stages of urban development following examples from the USA and Canada (Berry 1976; Bourne 1978; Bourne/Simmons 1978). They consider that after two stages of growth and concentration, then comes the critical 'pivotal point', characteristic of the mature industrial-urban society (p. 184). It is characterized by a 'relative decentralisation', when growth in the suburban-exurban rings exceeds that of the central city. Finally comes the later stage of 'absolute decentralisation' when the central city starts to decline.

The empirical evidence from 215 urban areas and their surrounding regions in Europe shows that 62 % were in the second stage during the period 1960–1970. They then wonder if the European Urban system would follow a '*clean break*' like that observed in the USA and Canada or whether they must consider it as a '*wave motion*' for Europe rather than a clean break (Gordon 1980), meaning an increasing concentration except in the core (stages of growth with de-concentration).

Deploring the lack of clear result on Europe, compared with US and Canada, they explained it by two sets of factors:

- Some national uniformities induce some structural or proportionality effects inside each country;
- Particularities of regions create differential or local-factor effects for specific sectors.

In order to test these two effects, they applied another sophisticated technique: Shift-Share Analysis. Shift-Share isolates the structural component (SC) representing the effect of industrial structure (strictly, employment) at the start of the period from the differential component (DC) representing the residual, that is the effect of different rates of growth in an industry in that region compared to the national average growth. As urban quantitative geographers, we would like to testify here to the great ability of Peter Hall to introduce strong and new visions of urban processes based mainly on a very subtle mix of quantitative treatments and qualitative information. Being interested overall by understanding the factors influencing the dynamics of cities, quantitative techniques allowed him to develop an extensive, large and ambitious research agenda, leading to the discovery of some very deep specific processes that he managed to explain with qualitative information.

Once completed, the results were very satisfying and highlighted some very uneven stages of urban development for cities in Europe, particularly in a national context that remained very strong for each country. This national effect remained strong 25 years later when Peter Hall and Kathy Pain deplore in the POLYNET project that:

> The analysis is complicated by the existence of national boundaries. In fact, despite major advances in integration, Europe is still a system of separated nation states, with separated languages and cultures (Hall/Pain 2006: 7).

It is clear that cities do not follow the same stages or same trajectory in their development. However, what Hall was most likely trying to imply through these analyses was that there is no specific path dependency a city must follow and that transitions are not smooth. Indeed, evidence from the previous section highlights why some cities might be more advanced than others based on their technological sophistication and speed of uptake. What perhaps is more important than some very generalised model of urban transformation is the recognition that cities or networks of cities must go through some stages or meet some requirements before a subsequent stage can be considered. This may be interpreted as technological dependency rather than path dependency. Peter Hall knew this inherently and provided us with an excellent example in the '*Forces Shaping Urban Europe*' paper (1993). In trying to identify the potential impacts on the spatial structure of Urban Europe he identifies transport technology as a key driver of urban change. It should be no surprise given his love of trains, identified many places elsewhere in this volume, that while he talks about all kinds of transport technologies, he specifically points out a new urban form—'*garopolis*'—whereby towns and cities might emerge as a result of TGV (high-speed train) stations built at strategic locations at substantial distances from Paris, (greater than 200 km) but clearly within commuting time of Paris given the speeds possible (300 km/h). Even if these garopolis are already explained per se by other testimonies in this volume, we are more sensible to his vision of how these 'garopolis' and subsequent changes to the urban system were made possible not only when the technology is available but also when other urban steps or stages are in place. In this way he highlighted how an urban hierarchy as a whole might be transformed, with each place being the sum of its adaptations in a multi-staged process. In a guest lecture he gave to a first year undergraduate class of British geographers at Canterbury Christ Church University in 2010 he did something quite similar. Using examples those students could all understand, he crafted a crystal clear argument where he demonstrated how the morphology of London, its network of sub-centres, and emergence of its first suburbs were the direct consequence of the expanding London underground network.

7.5 Multi-scale and Multi-level Approaches to the Study of Cities

The Cities/Regions/Nations approach undertaken in this European Urban System study with Hay (1980) remained a constant in his work and he always paid attention to the contextualization with multi-scale approach. He explicitly recognised globalisation and the emergence of trade blocs (in particular the EU) as exogenous forces that would affect places at all levels within any urban hierarchy (Hall 1993). Similarly, he was quick to produce work once major change was evident. What is so significant about the 1993 paper is his very early consideration of the likely spatial impacts the fall of the Berlin Wall would have on Eastern Europe. While he said it would be "foolhardy to speculate" on the future he made some modest predictions. Looking back now, much of the essence of what he postulated is there for all to see—the growth of multinationals, movement towards EU membership, migration into larger cities and the emergence of new nodes in a European transport network. None of these visions would be possible without the ability to take a holistic view of things induced through a multi-scale approach to urban Europe.

More recently, and also in the context of European cities, the POLYNET project endeavoured to explore the association between information flows and polycentric development at a regional scale (Hall/Pain 2006). Here they not only adopt a multi-scale approach but also a multi-level one. Indeed, one needs to distinguish between a multi-scale approach, that allows one to understand the close relations between cities and states or regional/continental blocks, with a multi-level approach that considers the feed-back loop effects between the inner organization of cities (meso level) and their general competitiveness (macro-level) (Fig. 7.1).

This multi-level approach always remained subjacent in his oeuvre, especially when his analyses and interpretations shift from the micro level of actors' behaviours, to the urban functioning of communities, local activities and urban fabric

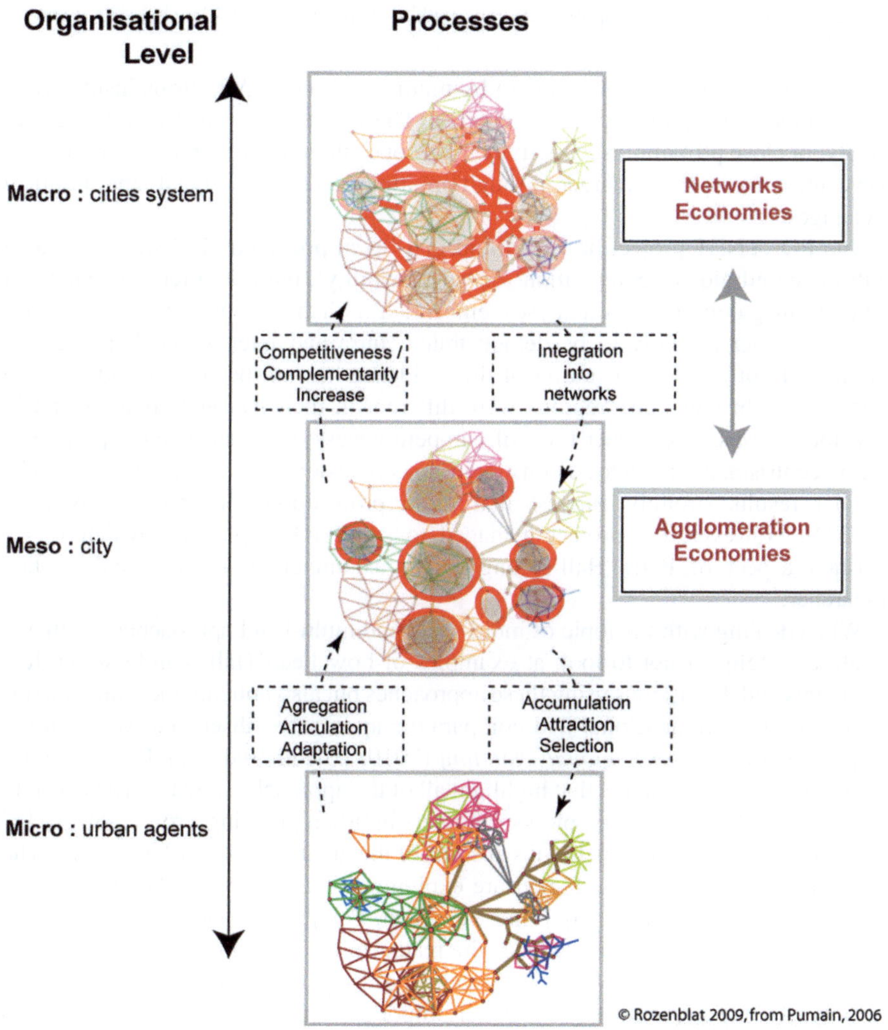

Fig. 7.1 Multi-level processes of urban systems. *Source* Rozenblat (2010)

(meso-level), to the national, continental or global competitiveness of cities (macro-level). But this multi-level framework appears much more explicitly in the POLYNET project (Hall/Pain 2006). Here, their main assumption claims that

> the degree of interrelationship among cities is a reflection of the concentration of advanced services within them; but conversely, this concentration reflects the degree of actual and potential connectivity between them; the process is circular and cumulative (Hall/Pain 2006: 7).

Hall/Pain (2006) concentrate on services beginning again to define them and determine their role in urban processes:

> All these are service industries processing information in a variety of ways (…) so strong agglomeration tendencies apply not only within each sector, but also between them (Hall/Pain 2006: 7).

They evaluated local polycentrism by multi-locations of APS firms inside urban areas, travels, telephone calls, internet… These activities interact all together locally, in close proximities, or activities are concentrated in different poles forming polycentric areas that interact through multiple economic, social and political exchanges.

The POLYNET project hoped to find that 'local polycentricity' was associated with increased global 'competition' or 'sustainability', but the differences amongst the eight mega-city regions analyses, embedded in their national settings, were too large for either property to be demonstrated. Intra and inter-urban dynamics are interlinked but the demonstration of this evidence has not been achieved here, so much the eight mega-city regions were different in their national contexts and in their local forms. The higher level of competitiveness or the sustainable properties of polycentrism, are neither demonstrated. In conclusion, they recognized that "The Polynet results strongly suggest that these propositions are not self-evident" (p. 209). However, the project managed to suggest key policies, which was a constant aspect of Peter Hall's long career as university lecturer and applied researcher.

When dealing with the topic of multi-scale and multi-level approaches to cities it would be negligent not to look at examples of how Peter Hall seamlessly crafted arguments and discussions using these approaches but also noticing the simultaneous use of the multi-dimensional and comparative approaches discussed earlier in the chapter. In *Urban and Regional Planning* (2010) co-authored with Tewdwr-Jones there are a series of chapters that highlight all of the approaches already described. In a chapter entitled "The seers: pioneer thinkers in urban planning, from 1880–1945" we are taken on a journey that looks at some of the great names in urban design. The work in of Howard, Unwin, Parker are explored in the context of 'Garden Cities', 'Letchworth' and the 'parkway principle'. Next, Perry, Stein and Tripp's contribution to planning and cities introduces the reader to the 'neighbourhood unit principle', 'Radburn', New Jersey and the 'precinctual' approach to traffic management in the city. Geddes recognised the wider importance of the 'natural region' and Abercrombie produced the 'Greater London Plan' of 1944, while Frank Lloyd

Wright presented us with 'Broad Acre City'. Moving away from those in the Anglo-American tradition, Hall and Tewdwr-Jones look to Europe for more developments in the urban realm. Soria y Mata's 'Linear City', Garnier's 'Cite industrielle', May's 'Trabantenstädte' (or satellite towns) and Le Corbusier's 'City of tomorrow' highlight alternative approaches to urban change in continental Europe.

What is truly inspirational here is the way in which these different icons of urbanism all worked at different scales—ranging from the street, to the precinct, to the neighbourhood, to the Garden City and New Towns, to suburban sprawl and finally to the metropolis and wider city region. It is with ease that such disparate scales and complex ideas are all brought together in a seemingly simple and straightforward manner—perhaps this was the secret to Peter Hall's success—the ability to distil vast amounts of information and knowledge into a straightforward, jargon-free, message.

In a similar vein to the use of multiple scales in the analysis of the principles of urban planning a chapter of the same book of Hall/Tewdwr-Jones (2010) is entitled, 'Planning in Western Europe since 1945'. Here Peter Hall highlights a multi-scalar, multi-level approach to the study of urban places, looking at all of Europe on a continental scale drawing on data from across the whole European Union in 2006 (Hall/Tewdwr-Jones 2010: 171). The focus then shifts to what would essentially be recognised by most as the EC12 scale (or Western Europe) where the symbolic imagery of 'Blue Banana' and 'Pentagon' as core regions of European dynamism are explored (p. 175). The chapter then turns its attention to France, its administrative and transport systems before zooming into focus on the Paris region and ultimately districts within the city, e.g. La Defense. The chapter continues by looking at Germany on a national scale and the Ruhr region, the Mezzogiorno region of Southern Italy, Scandinavian city-region planning with examples from Copenhagen's Regional Plan right down to the suburban/neighbourhood planning level in Stockholm, with the final destination of the chapter being the Randstad Holland Conurbation. The final section of the chapter tries to draw all the lessons learned from a huge range of processes (multi-dimensional or multi-level approach) at all geographical scales (multi-scale approach) across all of Europe (comparative approach).

Virtually the whole of Urban Europe is explored in the final section of the chapter, which updates and reconsiders much of Hall's earlier work on urban Europe as well as aspects of his later work on polycentricity. It is the dexterity with which so many places and so many ideas are synthesized, always with an eye on the policy perspective that is the hallmark of so much of Peter Hall's work. It is a skill which seems lacking in so many of today's younger researchers who tend to get too intimately involved with individual places at the expense of breadth of knowledge and relativity compared to other places.

What marks urban Europe today is increasing competition between all places up and down of the urban hierarchy and this often played out in the way cities promote themselves. Ironically, almost every city uses similar strategies to try to establish themselves as 'different'. Hall (1993) identified this promotion and marketing of cities, also known as 'boosterism', in Europe as a future source of competition between cities. He wrote of cities bidding to host major events, e.g. the Olympic Games, in the early 1990s in the hope of gaining competitive advantage. How appropriate when one considers it is unlikely London would ever have been awarded the 2012 Olympic Games were it not for ideas and policies promoted by Peter Hall in the 1970s and 1980s that led to the regeneration and transformation of the East End of London over the past 35 years.

7.6 Conclusion

The research Peter Hall produced whether in collaborations or on his own, has always remained faithful to clear objectives: understanding the driving forces behind urban development and being able to make predictions on possible futures for cities. The methods he developed had to be very strong and rigorous to tackle an organism as complex as a city with all its dimensions. His methods had to be suited to local scale analysis of the internal structures of individual cities yet flexible enough to explore the thousands of cities and associated data in a systemic analysis of a global hierarchy of urban systems. The methods had to be multi-dimensional taking into account linkages between different economic activities, between social and economic dimensions, between infrastructures and identities, between growth and sustainability. In a way he really contributed to the development of geography as a real 'complex science' even before 'complex science' was recognized at an institutional level. But most importantly his contributions helped decision makers at national and international level to understand better the urban phenomena. Moreover, he defined with a great honesty and optimism the spatial conditions for actions in cities, evaluating the 'degrees of freedom' required for planners and policies actions, while recognising the limitations: "The planner's range of possibilities is limited" (Hall 2000: 241).

As urban quantitative geographers, we would like to give testimony to Peter Hall's proven ability to provide strong and evocative new visions of urban processes of transformation, portraying complexities in such a way as to make them easily understandable. While much of his research was based on some fairly heavy-duty data gathering and statistical analysis his concise use of quantitative treatments and subtle thought provoking qualitative information, along with the excellent writing style, never left the impression you had read something other than interesting and inspirational. His research was often applied, always ambitious yet unassuming and he has done as much as anyone in the field to advance our knowledge of cities. If ever a British geographer and planner deserved a Knighthood it was Sir Peter Hall. He truly was a visionary.

References

Berry B.J.L. (1976) The Counterurbanization process: Urban America since 1970. In Berry (ed.) Urbanization and counterurbanization, *Urban Affairs Annual Reviews*, 11, Beverley Hills and London: Sage.

Bourne L. S. (1978) Emergent realities of Urbanization in Canada: Some parameters and Implications of declining Growth. Toronto, Canada, *Center for Urban and Community Studies, Research paper n°96*, University of Toronto.

Bourne L and Simmons J. (1978) *Systems of Cities: readings on structure, growth and policy*. New York, Oxford University Press.

Brotchie J.F., Hall P. and Newton P.W. (eds.) (1987) *The spatial impact of technological change*. Beckenham, Kent, UK, Croom Helm Ltd.

Castells M. and Hall P (1994) *Technopoles of the World*. London, Routledge.

Gordon P. (1980) Deconcentration without a "clean break". in Kawashima (ed.) *Urbanization processes: Experiences of Western and Eastern countries*. Oxford IIASA - Pergamon

Hall P. (1977). *The World cities*, second edition (1966 for the first edition). London, Weidenfeld and Nicholson.

Hall P. (1985) Geography of the Fifth Kondratieff. in Hall, P and Markusen, A. (eds.) (1985) *Silicon Landscapes*. London, Unwin Hyman, 1–19

Hall P. (1987a) The geography of the post-industrial economy. In Brotchie J.F., Hall P. and Newton P.W. (eds.) *The spatial impact of technological change*. Beckenham, Kent, UK, Croom Helm Ltd, 3–27

Hall P. (1987b) The geography of High Technology: an Anglo-American comparison. In Brotchie J.F., Hall P. and Newton P.W (eds.) *The spatial impact of technological change*. Beckenham, Kent, UK, Croom Helm Ltd, 141–158

Hall P. (1993) Forces Shaping Urban Europe. *Urban Studies*, 30, 6, 883–898

Hall P. (1998) *Cities in Civilization*. London, Weidenfeld and Nicholson.

Hall P. and Hay D. (1980) *Growth centres in the European urban system*. London, Heineman Educational Books.

Hall P. and Pain K. (ed.) (2006) *The Polycentric Metropolis: learning from mega-city regions in Europe*. London, Earthscan.

Hall P. and Pfeiffer U. (2000) *Urban future 21: A global agenda for twenty first century cities*. The Federal Ministry of Transport, Building and Housing of the Republic of Germany. London, E & FN Spon.

Hall P. and Tewdwr-Jones M. (2010) *Urban and regional planning*. London, Routledge.

Rozenblat C. (2010) Opening the black box of agglomeration economies for measuring cities' competitiveness through international firms networks. *Urban Studies*, November, 47(13) 2841–2865

Sassen S. (1991) *The Global City: New York. London, Tokyo*. Princeton, NJ, USA, Princeton University Press.

Chapter 8
Peter Hall's Publications

Richard D. Knowles

8.1 Books (Authored, Co-authored or Edited)

1962 *The Industries of London*. (Hutchinson, London).
1963 *London 2000*. (Faber, London).
1964 *Labour's New Frontiers*. (Ed.) (Deutsch, London).
1965 *Land Values*. (Ed.) (Sweet and Maxwell, London).
Londra nel 2000. (Marsilio Editori, Padova).
1966 *The World Cities*. (Weidenfeld, London). French, German, Italian, Spanish, and Swedish translations published simultaneously.
Von Thünen's Isolated State. (Ed.) (Pergamon Press, Oxford).
1967 *An Advanced Regional Geography of Northern and Western Europe*. With R.J. Harrison Church, G.R.P. Lawrence, W.R. Mead, and Alice Mutton (London, Hulton)
1969 *London 2000*. Second edition. (Faber and Faber, London).
1970 *Theory and Practice of Regional Planning*. (Pemberton, London).
1973 *The Containment of Urban England*. With R. Drewett, H. Gracey, R. Thomas (Allen and Unwin, London).
Planning and Urban Growth: An Anglo-American Comparison. With M. Clawson (Johns Hopkins, Baltimore).
1974 *Urban and Regional Planning*. (Penguin, Harmondsworth and David and Charles, Newton Abbot).
The Penguin World Atlas. (Ed.) (Penguin, Harmondsworth).
1977 *Europe 2000*. (Duckworth, London).
The World Cities. Second edition. (Weidenfeld and Nicolson, London).

Richard D. Knowles, Professor, University of Salford, UK, Emeritus Professor of Transport Geography; r.d.knowles@salford.ac.uk. This bibliography was compiled by Richard D. Knowles from Peter Hall's C.V., updated and corrected.

R.D. Knowles and C. Rozenblat (eds.), *Sir Peter Hall: Pioneer in Regional Planning, Transport and Urban Geography*, SpringerBriefs on Pioneers in Science and Practice 52, DOI 10.1007/978-3-319-28056-1_8

1978 *Issues in Urban Society*. (Ed., with R. Davies). (Penguin, London).
1980 *Great Planning Disasters*. (Weidenfeld, London). Paperback edition, 1981 (Penguin, London).
Growth Centres in the European Urban System. With D. Hay (Heinemann, London).
The Penguin World Atlas. Second edition. (Penguin, London).
1981 *The Inner City in Context*. (ed.) (Heinemann, London).
Transport and Public Policy Planning. (Ed., with D. Banister). (Mansell, London).
1982 *The Penguin World Travel Guide*. With M. Hall. (Penguin, London).
The World Cities. Chinese edition. (Progress Publishers, Peking).
Europe 2000. Chinese edition. (Academia Sinica, Nanking).
Urban and Regional Planning. Second edition. (Penguin, London).
Great Planning Disasters. American edition, with new introduction. (University of California Press, Berkeley and Los Angeles).
1984 *The World Cities*. Third edition. (Weidenfeld and Nicolson, London).
1985 *The Future of Urban Form: The Impact of New Technology*. (Ed., with J. Brotchie, P. Newton, P. Nijkamp). (Croom Helm, London and Sydney).
Silicon Landscapes. (Ed., with A. Markusen). (Allen and Unwin, Boston).
Can Rail save the City? The Impacts of Rail Rapid Transit and Pedestrianisation on British and German Cities. With C. Hass-Klau. (Gower, Aldershot).
1986 *High Tech America: The What, How, Where and Why of the Sunrise Industries*. With A. Markusen and A. Glasmeier (Allen and Unwin, Boston).
1987 *Western Sunrise: Genesis and Growth of Britain's High Tech Corridor*. With M. Breheny, D. Hart and R. McQuaid (Allen and Unwin, London).
The Spatial Impact of Technological Change. (Ed., with J. Brotchie and P. Newton). (Croom Helm, London).
1988 *The Carrier Wave: New Information Technology and the Geography of Innovation, 1846–2003*. With P. Preston (Unwin Hyman, London).
Cities of Tomorrow: An Intellectual History of Urban Planning and Design in the Twentieth Century. (Basil Blackwell, Oxford).
1989 *London 2001*. (Unwin Hyman, London).
1991 *The Rise of the Gunbelt: The Military Remapping of Industrial America*. With A. Markusen, S. Campbell and S. Deitrick (Oxford U.P., New York).
Cities of the 21st Century: New Technologies and Spatial Systems. (Ed., with J. Brotchie, M. Batty and P. Newton). (Melbourne: Longman Cheshire).
1993 *Multilateral Cooperation for Development in the Twenty-First Century: Training and Research for Regional Development*. (Ed., with R. de Guzman, C.M. Madduma Bandara, A. Kato). Nagoya: United Nations Centre for Regional Development.

1994 *Technopoles of the World: The Making of 21st-Century Industrial Complexes*. (With M. Castells). London: Routledge.
Las Tecnópolis del Mundo: La Formación de los Complejos Industriales del Siglo XXI. (With M. Castells). Madrid: Alianza Editorial.

1995 *Cities in Competition: Productive and Sustainable Cities for the 21st Century*. (Ed., with J. Brotchie, M. Batty, E. Blakely, and P. Newton. Melbourne: Longman Australia.

1996 *Four World Cities: a Comparative Study of London, Paris, New York and Tokyo*. London: Llewelyn Davies. (By Llewelyn-Davies, UCL Bartlett School of Planning and Comedia; Projector Director and Co-Author).
The People—Where Will They Go? National Report of the TCPA Regional Inquiry into Housing Need and Provision in England. (Ed., with M. Breheny). London: Town and Country Planning Association.
Urban and Regional Planning. Chinese edition. Taipei: Chu Liu Book Company.
Ciudades del Mañana: Historio del Urbanismo en el Siglo XX. Barcelona: Ediciones del Serbal.

1998 *Sociable Cities: The Legacy of Ebenezer Howard* (with C. Ward). Chichester: Wiley.
Cities in Civilization: Culture, Technology and Urban Order. London: Weidenfeld and Nicolson.
East West Perspectives on 21st Century Urban Development: Sustainable Eastern and Western Cities in the New Millennium. (ed. with Brotchie, J., Newton, P., Dickey, J.) Aldershot: Ashgate.

1999 *Sustainable Cities or Town Cramming?* London: Town and Country Planning Association.

2000 *Urban 21: World Report on the Urban Future 21* (with Pfeiffer, U.). Berlin: Federal Ministry of Transport, Building and Housing.
Urban Future 21: A Global Agenda for Twenty-First Century Cities. (with Pfeiffer, U.) London: Spon.
Urban 21: Der Expertbericht zur Zukunft der Städte. (with Pfeiffer, U.) Stuttgart: Deustche Verlags-Anstalt.

2002 *Cities of Tomorrow*. Third edition. Oxford: Blackwell.
Urban and Regional Planning. Fourth edition. London: Routledge.
Working Capital: Life and Labour in Contemporary London. (with Buck, N., Gordon, I., Harloe, M., Kleinman, M), London: Routledge.

2003 *To-Morrow: A Peaceful Path to Real Reform*. Edited, with an introduction, by Hall, P., Hardy, D. and Ward, C. London: Routledge.

2004 *Shrinking to Grow? The Urban Regeneration Challenge in Leipzig and Manchester*. (With Mace, A., Gallent, N., Porsch, L., Braun, R., Pfeiffer, U.) (2004) London: Institute of Community Studies.

2006 *The Polycentric Metropolis: Learning from Mega-City Regions in Europe* (with Pain, K.). London: Earthscan.

2007 *London Voices London Lives: Tales from a Working Capital*. Bristol: Policy Press.
2008 *Urban and Regional Planning*. Chinese Edition. Beijing: China Architecture and Building Press.
2011 *Urban and Regional Planning*. (with Tewdwr-Jones M.) Fifth Edition. London: Routledge. ISBN 978-0-415-56652-0 (hbk.)
2014 *Good Cities, Better Lives: How Europe Discovered the Lost Art of Urbanism*. London, Routledge.
Sociable Cities: The 21st Century Reinvention of the Garden City. (with Ward C.) London, Routledge

8.2 Contributions to Books

1964 The Development of Communications, and Industrial London: a General View. In Coppock, J.T. and Prince, H. (eds.), *Greater London,* 52–79. London: Faber and Faber.
General Introduction to the Excursions in Central London. In: Clayton, K.M. (ed.), *Guide to London Excursions.* London: Philip.
1965 Land Use and Development. In: Institution of Civil Engineers, *Conference on Transportation Engineering*. London: I.C.E.
1966 Urbanisation and the British Countryside: Note Towards a Research Project. In: Darling, F.F. and Milton, J.P. (eds.), *Future Environments of North America.* New York: Natural History Press.
1967 London. In: *Encyclopedia Britannica.* London: Encyclopedia Britannica Publications.
1968 The Urban Culture and the Suburban Culture. In: Eels, R., Walton, C. (eds.), *Man in the City of the Future.* New York: Collier-Macmillan.
1970 Transportation. In: Cowan, P. (ed.), *Developing Patterns of Urbanisation.* Edinburgh: Oliver and Boyd.
Regional Planning and Airport Location. In: Institution of Civil Engineers, *World Airports: The Way Ahead.* London: Evans.
1971 Britain 2000: Speculations on a Future Society. In: Chisholm, M. (ed.), *Resources for Britain's Future.* Newton Abbot: David and Charles; London: Penguin.
Spatial Structure of Metropolitan England and Wales. In: Chisholm, M., Manners, G. (eds.), *Spatial Policy Problems of the British Economy.* Cambridge: Cambridge U.P.
1972 The Pattern of Cities to Come. In: Barker, P. (ed.), *One for Sorrow, Two for Joy.* London: Allen and Unwin.
Planning Environment. In: Townsend, P. and Bosanquet, N. (eds.), *Labour and Inequality.* London: Fabian Society.

1973 Manpower and Education. In: Cowan, P. (ed.), *The Future of Planning.* London: Heinemann.

1974 England *circa* 1900. In: Darby, H.C. (ed.), *A New Historical Geography of England.* Cambridge: Cambridge U.P.

1975 Geography. In: Barker, P. (ed.), *The Social Sciences Today.* London: Edward Arnold.

1976 A Review of Policy Alternatives. In: Kehoe, D., Morley, D., Proudfoot, S., Roberts, N.A. (ed.), *Public Land Ownership: Frameworks for Evaluation.* Lexington: D.C. Heath.

1977 The Declining Metropolis: Patterns, Problems and Policies in Mainland Europe (with D. Metcalf.) In: Levin, C.W. *The Mature Metropolis.* Lexington: Lexington Books.

1978 The Geographer's View of Planning. In: Goodall, B., Kirby, A. (eds.), *Resources and Planning.* Oxford: Pergamon.

The European Experience in a National Policy for Balanced Urban Growth. In: *The White House Conference on Balanced National Growth and Economic Development.* Final Report, Vol. 3, Appendix. Washington, D.C.: Government Printing Office.

1979 Models of Development Strategy: A Critique. In: Rho, Y.-H., Hwang, M.-C. (eds.), *Metropolitan Planning: Issues and Policies.* Seoul: Korea Research Institute for Human Settlements.

1980 Technological Implementation Problems in the Development Process. In: *Proceedings of the 7th EAROPH Congress, Kuala Lumpur.* Kuala Lumpur: The Malaysian Organisation for Human Settlements.

Transportation. In: Cameron, G. (ed.), *The Future of the Conurbations.* London: Longman.

1981 Urban Change in Europe. In: Pred, A. (ed.) *Space, Time and Geography: Essays Dedicated to Torsten Hägerstrand.* Lund: CWK Gleerup.

1983 The Contribution of Scientific Research on Regional Affairs to Regional Policy-Making: The Case of the United Kingdom. In: *Research and Policy-Making: The Case of Regional Policy.* Paris: Organisation for Economic Cooperation and Development.

Decentralization without End? A Re-evaluation. In: Patten, J.A. (ed.) *The Expanding City: Essays in Honour of Jean Gottmann.* London: Academic Press.

Have Cities a Future? In: *Social and Urban Change: Its Impact on the Building Industry.* (First National Conference, Building Science Forum of Australia, New South Wales Division, Proceedings). Sydney: Building Science Forum.

Changing Development Hierarchies in the Development Process: An International Comparison. *International Geographical Union, Working Group on the Great Metropolitan Cities, Proceedings.* Rio de Janeiro: International Geographical Union.

Housing, Planning, Land and Local Finance: The British Experience. In: Lefcoe, G. (ed.) *Urban Land Policy for the 1980s*. Lexington, Mass.: D.C. Heath. Reprinted, *Urban Law and Policy, 6*, 1983, 75–86.

1984 Metropolis 1890–1940: Challenges and Responses. In: Sutcliffe, A. (ed.), *Metropolis 1890–1940*. London: Mansell; Chicago: University of Chicago Press.

Postscript: Metropolis 1940–1990. In: *Ibid.*, 431–445.

Geography: Descriptive, Scientific, Subjective and Radical Images of the City. In: Hollister, R., Rodwin, L. (ed.) *Cities of the Mind: Images and Themes of the City in the Social Sciences*. New York: Plenum Press.

Utopian Thought: A Framework for Social, Economic and Physical Planning. In: Alexander, P., Gill, R. (ed.) *Utopias*. London: Duckworth.

Ebenezer Howard. In: Barker, P. (ed.) *Founders of the Welfare State*. London: Heinemann.

The New Geography of Innovation. In: Bryant, C. (ed.) *Waterloo Essays in Geography*, Vol. 1, *Regional Economic Development*. Waterloo, Ontario: University of Waterloo.

The Urban Culture and the Suburban Culture: A New Look at an Old Paper. In: Agnew, J.A., Mercer, J., Sopher, D.E. (ed.) *The City in Cultural Context*, 120–133. Boston: Allen and Unwin.

1985 The World and Europe. In: *The Future of Urban Form: The Impact of new Technology*. (ed. with Brotchie, J., Newton, P., Nijkamp, P.) London and Sydney: Croom Helm,

Optimism and Pessimism in future Planning. In: *Ibid.*

The Geography of the Fifth Kondratieff. In: *Silicon Landscapes*. (ed. with A. Markusen). Boston: George Allen and Unwin.

The American Computer Software Industry: Economic Development Prospects (with A.R. Markusen, R. Osborn and B. Wachsman). In: *Ibid.*

High Technology and Regional-urban Policy (with A.R. Markusen). In: *Ibid.*

Dutch Urban Planning within the Perspective of European Development. In: Dutt, A.K., Costa, F.J. (ed.) *Public Planning in the Netherlands: Perspectives and Change since the Second World War*. Oxford: Oxford U.P.

Urban Transportation: Paradoxes for the 1980s. In: Jansen, G.R.M., Nijkamp, P., Ruijgrok C.J. (ed.) *Transportation and Mobility in an Era of Transition*. Amsterdam: North-Holland.

Common Problems—Different Solutions? In: *Strategies for Urban Revival: Report of 8th International INTA Conference, 21–28 October 1984*. The Hague: International New Towns Association.

Grosse Planungskatastrophen. In: Einem, E. von (ed.) *Die Rettung der kaputten Stadt*. Berlin: Transit.

1986 The State of Cities and the State of Planning. In: Hutchinson, B., Batty, M. (ed.) *Advances in Urban Systems Modelling*. Amsterdam: North-Holland.

Poblacio'n y crisis urbana. In: Olano, A. (ed.) *Tendencia's demograficas y Planifacio'n economica.* Madrid: Ministerio de Economi'a y Hacienda.

National Capitals, World Cities and the New Division of Labour. In: Ewers, H.-J., Goddard, J.B., Matzerath, H. (ed.) *The Future of the Metropolis: Berlin-London-Paris-New York: Economic Aspects.* Berlin: de Gruyter.

1987 Metropolitan Settlement Strategies. In: Rodwin, L. (ed.), *Shelter, Settlement and Development.* London: Allen and Unwin.

Perspectives on Post-Industrial Society and The Geography of High-Technology Industry: An Anglo-American Comparison. In Brotchie, J., Hall, P., Newton, P. (ed.) *The Spatial Impact of Technological Change.* London: Croom Helm.

La Metropoli Sovraffollata e il suo Centro Storico. In: Studi Centro storico Napoli (ed.) *Il Regno del Possibile.* Naples: Edizione 24 Ore del Sole.

1988 Urban Growth and Decline in Western Europe. In: Dogan, M., Kasarda, J. (ed.) *The Metropolis Era,* Vol. I, *A World of Giant Cities.* Newbury Park and London: Sage.

The Intellectual History of Long Waves. In: Young, M., Schuller, T. (ed.) *The Rhythms of Society.* London and New York: Routledge.

The Creation of the American Aerospace Complex, 1955–65: A Study in Industrial Inertia. In: Breheny, M. (ed.) *Defence Expenditure and Regional Development.* London and New York: Mansell.

Regions in the Transition to the Information Economy. In: Sternlieb, G., Hughes, J.W. (ed.) *America's New Market Geography: Nation, Region and Metropolis,* 137–160. New Brunswick: Rutgers University, Center for Urban Policy Research.

Introduction (with C. Hass-Klau). In: Hass-Klau, C. (ed.) *New Life for City Centres: Transport and Conservation in British and German Cities.* London: Anglo-German Foundation.

1989 The Rise and Fall of Great Cities: Economic Forces and Population Responses. In: Lawton, R. (ed.) *The Rise and Fall of Great Cities: Aspects of Urbanization in the Western World.* London and New York: Belhaven Press.

London 2001. In: Breheny, M., Congdon, P. (ed.) *Growth and Change in a Core Region: The Case of South East England. (London Papers in Regional Science, 20).* London: Pion.

Regional Development and Planning. (With R. Bromley, A. Dutt, D. Mookherjee, J.E. Benhart). In: Gaile, G.L., Willmott, C.J. (ed.) *Geography in America.* Columbus, Ohio: Merrill.

1990 High Technology Industry and the European Scene. In: *Urban Challenges,* 117–133. Stockholm: Statens offentliga utredningar, 1990:33.

Europe after 1992. In: *Urban Challenges,* 179–185. Stockholm: Statens offentliga utredningar, 1990:33.

Gibt es sie noch - die Stadt? In: Schabert, T. (ed.) *Die Welt der Stadt,* 17–42. Munich: Piper.

1991 The British Enterprise Zones. In: Green, R.E. (ed.) *Enterprise Zones: New Directions in Economic Development.* Newbury Park and London: Sage.

The Pentagon and the Gunbelt. In: Kirby, A. (ed.) *The Pentagon and the Cities* (*Urban Affairs Annual Reviews, 40*), 53–76. Newbury Park and London: Sage.

Metropolens Framtid. In: *The Soul and Form of the City* (*Framtider, 10/3*), 10–13.

Managing Growth in the World's Cities. In: Canadian Urban Institute, *Planning the Toronto Region: Lessons from Other Places: Proceedings* (*Urban Focus Series, 90–9*), 5–19. Toronto: Canadian Urban Institute.

Anonimia e Identità nella Supermetropoli. In: Nicolin, P. (ed.). *Atlante Metropolitano* 19–25. Milan: Electra.

The Restructuring of Urban Economics: Integrating Urban and Sectoral Policies. In: Fox-Przeworski, J., Goddard, J., de Jong, M. (ed.) *Urban Regeneration in a Changing Economy: The International Perspective*, 5–23. Oxford: Clarendon Press.

Keynote Address. In: Bayliss, D. (ed.) *Orbital Motorways: Proceedings of the Conference Organized by the Institution of Civil Engineers and held in Stratford upon Avon on 24–26 April 1990.* London: Thomas Telford.

Structural Transformation in the Regions of the United Kingdom. In: Rodwin, L. and Sazanami, H. (ed.) *Industrial Change and Regional Economic Transformation: The Experience of Western Europe*, 39–69. London: Unwin Hyman.

Europe's Regional-Urban Futures. (With R. Camagni, P. Cheshire, J.-P. de Gaudemar, L. Rodwin and F. Snickars). In: Rodwin, L. and Sazanami, H. (ed.) *Industrial Change and Regional Economic Transformation: The Experience of Western Europe*, 301–318. London: Harper Collins.

Moving Information: A Tale of Four Technologies. In: Brotchie, J., Batty, M., Hall, P., Newton, P. (ed.) *Cities of the 21st Century.* London: Routledge.

1992 New Technologies, Participation, Integration and Lifestyle. In: Organisation for Economic Co-operation and Development, *Cities and New Technologies*, 255–262. Paris: OECD.

The Pentagon and the Gunbelt (with A. Markusen). In: Kirby, A. (ed.) *The Pentagon and the Cities* (*Urban Affairs Annual Reviews, 40*), 53–76. Newbury Park and London: Sage.

New Technologies, Participation, Integration and Lifestyle. In: Organisation for Economic Co-operation and Development, *Cities and New Technologies*, 255–262. Paris: OECD.

Transport: Maker and Breaker of Cities. In: Mannion, A.M., Bowlby, S.R. (ed.) *Environmental Issues in the 1990s*, 265–276. Chichester: John Wiley.

1993 Priorities in Urban and Economic Development. In: Summers, A.A., Cheshire, P.C., Senn, L. (ed.) *Urban Change in the United States and Western Europe: Comparative Analysis and Policy*, 55–85. Washington, D.C.: The Urban Institute Press. (Second Edition: 1999).

Comments on Dutch Planning. In: Faludi, A. (ed.) *Dutch Strategic Planning in International Perspective*, 37–45. Amsterdam: Netherlands Universities Institute for Coordination of Research in Social Sciences (SISWO Publication 372).

Cities and Regions in a Global Economy. In: Hall, P., Guzman, R. de, Madduma Bandara, C.M., Kato, A. (ed.) *Multilateral Cooperation for Development in the Twenty-First Century: Training and Research for Regional Development*, 6–26. Nagoya: United Nations Centre for Regional Development.

Der Einfluss des Verkehrs und der Kommunikationstechnik auf Form und Funktion der Stadt. In: Finke, L. et al., *Zukunft Stadt 2000: Stand und Perspektiven der Stadtentwicklung*, 383–449. Stuttgart: Deutsche Verlags-Anstalt.

Red Viaria Principal y Desarrollo Urbano. In: Anon: *Movilidad y Territorio en las grandes Ciudades: El Papel de la Red Viaria*. Madrid: Communidad de Madrid, Consejería de Transportes and Spain, Ministerio de Obras Públicas, Transportes y Medio Ambiente, Dirección General de Planificación Intermodal del Transporte en las Grandes Ciudades.

The Changing Role of Capital Cities: Six Types of Capital City. In: Taylor, J., Lengellé, J.G., Andrew, C. (ed.) *Capital Cities—Les Capitales: Perspectives Internationales—International Perspectives*, 69–84. Ottawa: Carleton U.P.

1994 The Disappearing City. In: Cuadrado-Roura, J.R., Nijkamp, P., Salva, P. (ed.) *Moving Frontiers: Economic Restructuring, Regional Development and Emerging Networks*, 23–35. Aldershot: Avebury.

London 1994: Retrospect and Prospect. In: Simmie, J. (ed.) *Planning London*, 172–187. London: UCL Press.

Vision for the Future. [Report of the 2nd Annual Vision for London Lecture 1994]. London: Vision for London.

1995 Il Futuro della Metropoli e sua Forma. In: Boscacci, F., Camagni, R. (ed.) *Tra Città e Campagna*, 89–112. Milan: Il Mulino.

Towards a General Urban Theory. In: Brotchie, J., Batty, M., Blakely, E., Hall, P., Newton, P. (ed.) *Cities in Competition: Productive and Sustainable Cities for the 21st Century*, 3–31. Melbourne: Longman Australia.

The Future of Planning. In: Giersch, H. (ed.) *Urban Agglomeration and Economic Growth*, 257–272. Berlin: Springer.

The Roots of Urban Innovation: Culture, Technology and Urban Order. In: Anon, *Cities and the New Global Economy: Conference Proceedings Volume 1*, 275–293. Canberra: Australian Government Publishing Service.

A European Perspective on the Spatial Links between Land Use, Development and Transport. In: Banister, D. (ed.) *Transport and Urban Development*, 65–88. London: Spon.

Summary and Conclusions (with D. Banister). In: Banister, D. (ed.) *Transport and Urban Development*, 278–287. London: Spon.

Can Cities be Sustainable? In: Serageldin, I., Cohen, M.A., Sivaramakrishnan, K.C. (ed.) *The Human Face of the Urban Environment: Proceedings of the Second Annual World Bank Conference on Environmentally Sustainable Development*, 32–38. Washington, DC: The World Bank.

1996 London and Berlin in the Next Century: Where Will They Position Themselves? In: Aaronovitch, S. et al., *Unemployment and the Economy of the City: London and Berlin Compared*, 12–18. London: Anglo-German Foundation.

The European City—Past and Future. In: Denmark, Ministry of Environment and Energy, Spatial Planning Department (ed.) *The European City—Sustaining Urban Quality: A Working Conference, Copenhagen, 24–28 April 1995*. Part 1: *Keynote Speeches and Workshop Conclusions*, 18–27. Copenhagen: Miljø & Energi Ministeriet.

High-Technology Industry in the New York Metropolitan Area: A View from History. In: Raymond, S.U. (ed.) *The Technology Link to Economic Development: Past Lessons and Future Imperatives*. (*Annals of the New York Academy of Sciences, 787*), 42–66.

Strategic Metropolitan Planning: The London Experience. In: Bell, J.K., Webber, S. (ed.) *Urban Regions in a Global Context: Directions for the Greater Toronto Area. (Proceedings of a Conference, University of Toronto, October 18–20, 1995)*, 7–18. Toronto: University of Toronto, Centre for Urban and Community Studies and Program in Planning.

Creative Cities. In: Hardy, S., Malbon, B., Taverner, C. (ed.) *The Role of Art and Sport in Local and Regional Economic Development*, 107–113. (*Proceedings of the Regional Studies Association*, Annual Conference, November 1996). London: Regional Studies Association.

Foreword. In: Banham, R. *A Critic Writes: Essays by Reyner Banham*, xi–xv. Berkeley: University of California Press.

1997 Den Europeiska Staden - förr och i Framtiden. In: Nyström, L. (ed.) *Stadslandskapet: Sönderfall eller Läkning*, 13–24. Karlskrona: Stadsmiljörådet. (*Framtidsstaden*, IV).

Foreword. In: Banham, R. *A Critic Writes: Essays by Reyner Banham*, xi–xv. Berkeley: University of California Press.

The University and the City. In: University of Copenhagen, *Centre of Excellence*, 30–48. Copenhagen: University of Copenhagen.

The View from London Centre: Twenty-five Years of Planning at the DOE. In: Blowers, A., Evans, R. (ed.) *Town Planning into the 21st Century*, 119–136. London: Routledge.

Contributor to: Santer, J., Delors, J. (ed.) *La Commission Européenne à l'Écoute du Changement*. Rennes: Éditions Apogée (*Collection Politique Européenne*).

Sesso, Nevrosi e Impotenza Politica: la Trista Storia della Regional Planning Association of America. In: Ventura, F. (ed.) *Alle Radici della Città*

Contemporanea: Il Pensiero di Lewis Mumford, 133–154. Milan: CittàStudiEdizioni.

Reflections on Swedish Planning. In: Guinchard, C.G. (ed.) *Swedish Planning: Towards Sustainable Development*, 31–33. Gävle: Swedish Society for Town and Country Planning.

1998 Information Technology, Globalization and Regional Development. In: Gravesteijn, S.G.E., van Griensven, S., de Smidt, M.C. (ed.) *Timing Global Cities* (Nederlandse Geografische Studies 241), 101–115. Utrecht: Koninklijk Nederlands Aardrijkskunding Genootschap/Vereninging van Utrechtse Geografie Studenten VUGS.]

Changing Geographies: Technology and Income. In: Schön, D.A., Sanyal, B., Mitchell, W.J. (ed.) *High Technology and Low-Income Communities: Prospects for the Positive Use of Advanced Information Technology*, 43–68. Cambridge, Mass: M.I.T. Press.

Amendola, G. (1998) Interview with Peter Hall, Urban Planner. In: Council of Europe (1998) *Culture and Neighbourhoods: Perspectives and Keywords*, 39–57. Strasbourg: Council of Europe Publishing.

1999 Planning for the Mega-City: A New Eastern Asian Urban Form? In: Brotchie, J., Newton, P., Hall, P., Dickey, J. (ed.) *East West Perspectives on 21st Century Urban Development: Sustainable Eastern and Western Cities in the New Millennium*, 3–36. Aldershot: Ashgate.

Creative Cities and Economic Development. In: Anon, *World Urban Economic Development: The Official Business and Technology Briefing for the World Competitive Cities Congress, 19–21 May 1999, Washington, D.C.*, 24–27. London: World Markets Research Centre.

The Future Planning of City Regions. In: Gaffikin, F., Morrissey, M. (ed.) *City Visions: Imagining Place, Enfranchising People*, 61–78. London: Pluto Press.

The Regional Dimension. In: Cullingworth, J.B. (ed.) *British Planning: 50 Years of Urban and Regional Policy*, 76–90. London: Athlone.

Priorities in Urban and Economic Development. In: Summers, A.L., Cheshire, P.C., Senn, L. (ed.) *Urban Change in the United States and Western Europe: Comparative Analysis and Policy* (Second Edition), 57–88. Princeton: Princeton U.P.

Future Urban Lifestyles. In: Bundesamt für Bauwesen und Raumordnung, *Urban Future: Preparatory Expertises (Overviews) for the World Report on Urban Future for the Global Conference on the Urban Future URBAN 21*, 31–40. Berlin: Bundesamt für Bauwesen und Raumordnung (*Forschungen*, Heft 92).

The Creative City in the Third Millennium. In: Verwijnen, J., Lehtovuori, P. (ed.) *Creative Cities: Cultural Industries, Urban Development and the Information Society*, 36–57. Helsinki: University of Art and Design.

Redefining Europe's Cities. In: Bellet, C., Llop, J.M. (ed.) *Ciudades Intermedias: Urbanización y Sostenibilidad*, 93–108. Lleida: Milenio.

2000 Innovative Cities. In: Anon, *Structural Change in Europe: Innovative Cities and Regions*, 31–35. Bollschweil: Hagbarth.

The Centenary of Modern Planning. In: Freestone, R. (ed.) *Urban Planning in a Changing World: The Twentieth Century Experience*, 20–39. London: Spon.

The Evolution of Post-War Town Centres. In: Falk, N. (ed.) *The Renaissance of Post-War Town Centres*. London: Urbed.

2001 Global City-Regions in the Twenty-first Century. In: Scott, A.J. (ed.) *Global City-Regions: Trends, Theory, Policy*, 59–77. Oxford: Oxford U.P.

Urban Development and Research Needs in Europe. In: CERUM (Centre for Regional Science) *Urban Design for Sweden and Europe: Conference Report, June 20–21, 2000*. Umeå: CERUM.

2002 Introduction: London: Creative City. In: Pick, C. (ed.) *Dynamic Environments: London's New Entrepreneurs at Work*, 8–12. London: Workspace Group.

The Creativity of Cities: An Idea whose Time has Come. In: Netherlands, Deltametropolis Association, *Generating Culture: Roots and Fruits*, 30–45. Den Haag: Deltametropole/Ministerie van VROM (Forum).

Die Europäische Stadt, 2025: Dilemmata und Lösungen der Zukunft. In: Mayr, A., Meurer, M., Vogt, J. (ed.) *Stadt und Region; Dynamik von Lebenswelten*, 42–56. Leipzig: Deutschen Gesellschaft für Geographie.

Christaller for a Global Age: Redrawing the Urban Hierarchy. In: Mayr, A., Meurer, M., Vogt, J. (ed.) *Stadt und Region; Dynamik von Lebenswelten*, 110–128. Leipzig: Deutschen Gesellschaft für Geographie.

2003 Introduction, Commentary and Postscript. (with Hardy, D., Ward, C.) (2003) In: Howard, E., *To-Morrow: A Peaceful Path to Real Reform*. London: Routledge.

La Città Sostenibile in un'Epoca di Globalizzione. In: Girard, F.F., Forte, B., Cerreta, M., De Toro, P., Forte, F. (ed.) *L'Uomo e la Città: Verso uno Sviluppo Umano e Sostenibile*, 94–110. Milano: FrancoAngeli.

Geographers and the Urban Century. In: Johnson, R., Williams, M. (ed.) *A Century of British Geography*, 545–562. Oxford: Oxford U.P.

Smart Growth on Two Continents. In: Neal, P. (ed.) *Urban Villages: and the Making of Communities*, 33–47. London: Spon.

Australian Planning: Millennial Reflections. In: Toon, J., Falk, J. (ed.) *Sydney: Planning or Politics: Town Planning for the Sydney Region since 1945*, 15–26. Sydney: University of Sydney, Planning Research Centre.

The Sustainable City in an Age of Globalization. In: Fusco Girard, L., Forte, B., Cerreta, M., De Toro, P., Forte, F. (ed.) *The Human Sustainable City: Challenges and Perspectives from the Habitat Agenda*, 55–70. Aldershot: Ashgate.

2004 Where it's Happening: Innovation, Clustering and Dynamic Cities in the New EU. In: Anon, *Innovative City and Business Regions* (Structural Change in Europe, 3), 19–26. Bollschweil: Hagbarth.

The Changing European Urban System. In: Portas, N., Domingues, A., Cabral, J. (ed.) *Políticas Urbanas: Tendências, Estratégias e Oportunidades*, 226–235. Lisbon: Fundação Calouste Gulbenkian.

Londres: La Transformacion de una Ciudad (London: A City Transformed). In: Dirección General de Arquitectura y Vivienda (ed.) *Ciudad+Arquitectura: Londres (City+Architecture: London)*, 9–21. Sevilla: Egondi Artes Gráficas.

Learning from Manchester (with Mace, A.). In: Daldrup, E.L., Doehler-Behzadi, M. (ed.) Plus Minus Leipzig 2030: *Stadt in Transformation/Transforming the City*, 82–91. Wuppertal: Müller +Busmann.

Mega-Projects and Mini-Projects. In: Robinson, P., McCarthy, J., Forster, C. *Urban Reconstruction in the Developing World: Learning through International Best Practice*, 14–24. Sandown, SA: Heinemann.

London: Competitiveness, Cohesion and the Policy Environment. (with Gordon, I. et al.). In: Boddy, M., Parkinson, M. (ed.) *City Matters: Competitiveness, Cohesion and Urban Governance*, 71–90. Bristol: Policy Press.

European Cities in a Global World. In Eckardt, F., Hasenpflug, D. (ed.) *Urbanism and Globalization,* 31–46. Frankfurt: Peter Lang.

The Key to Creativity. In: Wood, P. (ed.) *Intercultural City Reader*, 26–36. (Intercultural City Series: Book 1). Stroud: Comedia.

Megaciudades, Ciudades Mundiales y Ciudades Globales. In: Ramos, A.M. (ed.) *Lo Urbano: en 20 Autores Contemporáneos*. Barcelona: Edicions UPC.

2005 Integrated Urban Transport and Urban Planning: as the Maker and Breaker of Cities (with Hickman, R.). In: Anon, *Entrepreneurial Spirit in Cities and Regions (Structural Change in Europe, 4)*, 86–92. Bollschweil: Hagbarth.

Retro Urbanism: On the Once and Future TOD. In: Saunders, W.S. (ed.) *Sprawl and Suburbia*, 122–129. Minneapolis: University of Minnesota Press.

2006 From Coronation to Jubilee. In: Buonfino, A., Mulgan, G. (ed.) *Porcupines in Winter: The Pleasures and Pains of Living Together in Modern Britain*, 16–22. London: Young Foundation.

Seven Types of Capital City. In: Gordon, D.L.A. (ed.) *Planning Twentieth Century Capital Cities,* 8–14. London: Routledge.

What is the Future of Capital Cities? In: Gordon, D.L.A. (ed.) *Planning Twentieth Century Capital Cities,* 270–274. London: Routledge.

Jean Gottmann (1915–1994). (With Johnston, R., Clout, H.) In: Armstrong, P.H., Martin, G.J. (ed.) *Geographers: Biobibliographical Studies, 25,* 42–59. London: Continuum.

London: A Millennium-Long Battle, A Millennial Truce? In: Tewdwr-Jones, M., Allmendinger, P. (ed.) *Territory, Identity and Spatial Planning: Spatial Governance in a Fragmented Nation*, 167–182. London: Routledge.

Central Lancashire: The Necessary Mythical Beast? In: Northwest Regional Development Agency and Centre for Cities at ippr (2006) *Cities Northwest*, 39–43. Warrington: NWDA.

The Land Fetish: Densities and London Planning. In: Kochan, B. (ed.) *London: Bigger and Better?*, 84–93. London: LSE London, London School of Economics and Political Science.

Why Some Cities Flourish while Others Languish. In: UN-Habitat, *The State of the World's Cities Report, 2006/2007*, 13. London: UN-Habitat/Earthscan.

Heathrow: A Retirement Plan. (With Hall, T.) London: Town and Country Planning Association.

2007 Planning and Sustainable Development of Metropolitan Capital Regions. In: Dirección General de Urbanismo y Planificacíon Regional. Consejeria de Medio Ambiente y Ordenación del Territorio (2007) *Capital Regions Conference: Congress Patron, Sir Peter Hall, Madrid, May 2006*, 25–47. Madrid: Communidad de Madrid.

The United Kingdom's Experience in Revitalizing Inner Cities. In: Ingram, G., Hong, Y.-H. (ed.) *Land Policies and their Outcomes: Proceedings of the 2006 Land Policy Conference*, 259–283. Cambridge, Mass: Lincoln Institute of Land Policy.

Delineating Urban Territories: Is this a Relevant Issue? In; Cattan, N. (ed.) *Cities and Networks in Europe: A Critical Approach of Polycentrism*, 3–14. Montrouge: Éditions John Libbey Eurotext.

Evolution of Strategic Planning and Regional Development. In: Dimitriou, H., Thompson, R. (ed.) *Strategic Planning for Regional Development in the U.K.: A Review of Principles and Practice*, 10–27. London: Routledge.

Europe's Multi-Centred Urban Regions. In: *Cities and Regions: Facing Up to Change* (*Structural Change in Europe, 5*), 24–26, 79. Bollschweil: Hagbarth Publications.

The Production of Affordable Housing in the U.K. (With Gallent, N.) In: (Anon) *A Tale of Two Cities: China-UK Comparative Study on Housing Provision for Low-Income Urban Residents*, 59–107. Guangdong: Cultural and Educational Section of the British Consulate-General.

Land, Shelter, Transport: The Latin American Way. In: Freire, M., Lima, R., Cira, D., Ferguson, B., Kessides, C., Mota, J.A., Motta, D. (ed.) *Land and Urban Policies for Poverty Reduction*, 21–33. Washington, DC: World Bank.

Measuring a Polycentric Europe: Whose Space of Flows? In: (Anon) *Geophilia: O Sentir e os Sentidos da Geografia: Homenagem a Jorge Gaspar*, 613–635. Lisbon: Centro de Estudos Geográficos da Universidade de Lisboa.

Preface: Land, Shelter, Transport: The Latin American Way. In: Freire, M. et al. (ed.) *Land and Urban Policies for Poverty Reduction: Proceedings of the Third International Urban Research Symposium held in Brasilia, April*

2005, 21–33. Brasilia: The World Bank and Institute for Applied Economic Research

Spaces and Flows in Contemporary Europe. In: Anon, *Demain la Ville/De Stad van Morgen/The City of Tomorrow: International Colloquium on the Future of Cities*, 109–124. Brussels: Regional Centre for Urban Development.

2009 Planning London—A Conversation with Peter Hall. In: Imrie, R., Lees, L., Raco, M. (ed.) *Regenerating London: Governance, Sustainability and Community in a Global City*, 24–39. London: Routledge. ISBN 13: 978-0-415-43366-2.

Geography and Planning: A New Version of an Old Story? In: Turan, N., Ramos, S. (ed.) *New Geographies 1: After Zero,* 146–154. Cambridge, Mass.: Harvard University Press. ISBN 978-1-934519-20-9.

Sociable Cities: The Legacy of Ebenezer Howard. (with Ward C.) Chinese edition. Beijing: China Architecture and Building Press. ISBN 978-7-112-10795-7.

Better Rail Stations: An Independent Review presented to Lord Adonis, Secretary of State for Transport. (with Green C.) London: Queen's Printer and Controller of Her Majesty's Stationery. No ISBN.

Top Ten Planning Ideas. In: Hack, G., Birch, E.L., Sedway, P.H., Silver, M.J. (ed.) *Local Planning: Contemporary Principles and Practice*, 165–172. Washington, DC: ICMA. (An ICMA Green Book). ISBN 978-0-87326-148-7.

L'Art de transformer les Métropoles au XXIe Siècle. (with Balducci, A, Leboreiro, Lecroart, P.) (*Les Cahiers de l'IAU île-de-France, 151,* 80–84. ISSN 0153-6184.

2010 Sin Plan: Un Experimento sobre la Libertad. (with Banham, R., Barker, Price, C.) In: Walker, E. (ed.) *Lo Ordinario,* 37–59. Barcelona: Editorial Gustave Gill. ISBN 978-84-252-2330-3.

The Polycentric Metropolis: A Western European Perspective on Mega-City Regions. In: Xu, J., Yeh, A.G.O. (ed.) *Governance and Planning of Mega-City Regions: An International Comparative Perspective*, 29–50. Abingdon: Routledge. ISBN 978-0-415-56089-4 (hbk), 978-0-203-86551-4 (ebk).

2011 Looking Backward, Looking Forward: The City Region of the Mid-21st Century. In: Neuman, M., Hull, A. (ed.) *The Futures of the City Region*, 15–29. London: Routledge. ISBN 13: 978-0-415-58803-4.

Regions and Regional Policy: a Global View. In: Richardson, H.W., Bae, C.-H.C., Choe, S.-C. (ed.) *Reshaping Regional Policy*, 21–40. Cheltenham: Edward Elgar. ISBN: 978-1-84980-280-2.

Acting or Reacting: How Cities, City Regions and Landscapes are changing the Urban System in Europe. In: Altrovk, U., Aring, J., Hahne, U., Reuther, I. (ed.) *Gewinnen Verlieren Transformieren: Die Europäischen Stadtregionen in Bewegung*, 189–204. Berlin: Reimer. ISBN 978-3-496-01436-2.

2012 Balancing European Territory: The Challenge of the Post-Neoliberal Era in Ganser R. and Piro R. (eds.) *Parallel Patterns of Shrinking Cities and Urban Growth: Spatial Planning for Sustainable Development of City Regions and Rural Areas.* 27–44. Farnham, UK: Ashgate

2014 Apologia Pro Vita Sua. In Tewdwr-Jones M., Phelps N., and Freestone R (eds.) *The Planning Imagination: Peter Hall and the Study of Urban and Regional Planning*. London, Routledge: 269–284.

2015 High-speed Trains and Spatio-Economic Impacts: A British-French Comparison on Two Scales: Intra- and Inter-regional. In Hickman R., Givoni M., Bonilla D. and Banister D. (eds.) *International Handbook on Transport and Development*. London: Edward Elgar: 301–317.

A History of Contemporary Planning: Personal Stories, Personal Journeys in Britain and America. In Haselberger B. and Saija L., (eds.) *Encounters in Planning Thought: Autobiographical Essays from Key Thinkers in Planning* (Forthcoming)

8.3 Periodicals: Articles and Research Journal Papers

1960 The Location of the Clothing Trades—East London, 1851–1951. *Transactions, Institute of British Geographers, 28*, 155–78.

1962 The East London Footwear Industry: An Industrial Quarter in Decline. *East London Papers, 5*, 3–22.

1967 New Techniques in Regional Planning: Experience of Transportation Studies. *Regional Studies, 1*, 17–21.

Regional Planung in London und Sudost England. *Bauwelt, 58*, 1216–19.

Planning for Urban Growth: Metropolitan Area Plans and Their Implications for South East England. *Regional Studies, 2*, 101–134.

1970 Transportation. *Urban Studies, 6*, 408–76. Subsequently published in P. Cowan (ed.), *Developing Patterns of Urbanization* (Oliver and Boyd, Edinburgh).

1972 Beyond the Fringe? *Design, 280*, 25–29 (with M. Marcus.)

1973 Forecasting the Quality of Life in Urban Europe. *Greater London Council IntelligenceUnit Quarterly Bulletin,* 5–16.

Regional and Urban Planning: A View from Great Britain. *Technology and Society, 8*, 82–84.

Urban Europe 2000: Progress of a Research Project. *Futures, 5*, 449–456.

1974 The Containment of Urban England. *Geographical Journal, 140*, 386–417.

The New Political Geography. *Transactions of the Institute of British Geographers, 63*, 48–52.

La estructura espacial de la economia espanola: su reciente desarrollo y planificacion future. *Revista Espanala de Economia, 413*, 147–170.

1977 Urban and Regional Planning in Britain and America: Ends and Means. *Planning Outlook, 20*, 19–22.
Urban Structure and Modal Split in the Journey to Work, with R. J. Sammons. *Urban Studies, 14*, 1–9.
1978 How to Build London's Ringways at Low Cost and Almost No Environmental Pain, with E. Smith and V. Evelyn. *The London Journal, 3*, 3–23.
The Roots of Urban Change. *The Ditchley Journal, 5/1*, 20–27.
Cities in Europe 2000. *Futures, 10*, 469–475.
Can Cities Survive?: The Potential and Limits of Planning. *The Ditchley Journal, 5/2*, 33–41.
1979 Private Enterprise in the Contemporary City. *The Ditchley Journal, 6/1*, 5–15.
1980 Great Planning Disasters. *Futures, 12*, 45–50.
Finance and Land Use in Urban Development. *The Ditchley Journal, 6/2*, 23–35.
New Trends in European Urbanization. *The Annals of the American Academy of Political and Social Science, 451*, 45–51.
Making Cities Liveable. *The Ditchley Journal, 7/2*, 24–37.
1981 The Life and Death of a Quasi-Quango: The South East Economic Planning Council. *The London Journal, 6*, 196–206.
The Geographer and Society. *Geographical Journal, 147*, 145–152.
1982 The new Political Geography: Seven Years On. *Political Geography Quarterly, 1*, 65–76.
Enterprise Zones: British Origins, American Adaptations. *Built Environment, 7*, 5–12.
Enterprise Zones: A Justification and Response. *International Journal of Urban and Regional Research, 6*, 416–421 and 443–445.
Regional Development in the EEC: A Look Back and a Look Forward. *Built Environment, 7*, 229–232.
Keys to Regional Growth. *Society, 19/5*, 48–52.
Ebenezer Howard. *Town and Country Planning, 52*, 1983, 42–47.
1983 Whither Regional Planning? *The Planner, 69*, 113–4.
The Anglo-American Connection: Rival Rationalities in Planning Theory and Practice, 1955–1980. *Environment and Planning, A, 5*, 41–46.
Land Use Change and Transport Policy. *Habitat International, 7*, 67–77.
Changing Urban Hierarchies in the Development Process: A Comparison. *Habitat International, 7*, 129–135.
1984 Have our Cities got a Future? *Town and Country Planning, 53*, 78–80.
Metropolitan High-Technology Industry in the Mid 1970s: Can Everyone have a Slice of the High-Tech Pie? (with A. Glasmeier and A. Markusen) *Berkeley Planning Journal, 1*, 131–142.
Have Cities a Future? *Futures, 16*, 344–350.

Enterprises of great Pith and Moment? *Town and Country Planning, 53*, 296–298.
Cities of the Future. In: Council of Europe, *Urban Renaissance in Europe: Study Series, 22*, 27–32.
Demographic Development and Revitalization of the City. *Toshi-Mondai Kenkyu*, 1–18. Osaka: Osaka University.
The Corridors of Change. *ESRC Newsletter, 53*, 20–21.

1985 21st Century Britain. *The Geographical Magazine, 57*, 1985, 228–230.
The Decline of the Cities: A Problem with its Roots in the distant Past. *Town and Country Planning, 54*, 40–43.
The People: Where Will they Go? (The first J.R. James Memorial Lecture). *The Planner, 71/4*, 3–12.
Revitalisation of Inner City Areas—A Summing Up. In: International Federation for Housing and Planning, *1984 Papers and Proceedings, 37th World Congress, Berlin (West)*, 97–100.
Winning back the City. *Australian Planner, 23/2*, 16–18.
Capitales nacionales, Ciudades internacionales y la nueva Division de Trabajo. *Estudios Territoriales, 19*, 21–30.

1986 Scenarios for Europe's Cities. *Futures, 18*, 2–8.
From the unsocial City to the social City. (The sixth J.R. James Memorial Lecture). *The Planner, 72/3*, 17–24.
Cities of the Aged. *Populi, 13/1*, 45–51.
Can Rail save the City?. *Plan Canada, 26/4*, 99–103 (with C. Hass-Klau).
Urban Actors: Puppets, Ad Libbers, or Something in Between? *Berkeley Planning Journal, 3*, 67–75 (with the members of CP 250).
Completing the grand Design. *Town and Country Planning, 55*, 233–235.
The New Zealand urban System: Deurbanisation at the southern Periphery of the Urban World. *New Zealand Geographer, 42*, 65–69.
Can Rail save the City?: The Impacts of Rail Rapid Transit on British and German. Cities, *UITP Review, 3/1986*, 203–217 (with C. Hass-Klau).

1987 Key to Success for Cities. *Town and Country Planning, 56*, 50–52 (with P. Cheshire).
The Anatomy of Job Creation: Nations, Regions and Cities in the 1960s and 1970s. *Regional Studies, 21*, 95–106.
Urban Development and the Future of Tourism. *Tourism Management, 8*, 129–130.
Las Ciudades de Europa: Un Problema Europeo? Una Profesion Europea? *Urbanismo, 1*, 25–31.
Great Opportunity in Central London. *Town and Country Planning, 56*, 236–7 (With N. Falk).
North-South Divide: Time for equal Ways to succeed. *Town and Country Planning, 56*, 323–325.

Biotechnology: The next Industrial Frontier. *Built Environment, 13*, 145–146.

Molecular Biology and Biotechnology: From a new Science to a new Industry. *Built Environment, 13*, 147–151.

Where Biotechnology locates. (With M.M. Webber, L. Bornstein, R. Grier). *Built Environment, 13*, 152–156.

1988 The Coming Revival of Town and Country Planning. *Journal of the Royal Society for the Encouragement of Arts, Manufactures and Commerce, 136*, 417–430. Also in *Town and Country Planning, 57*, 40–45.

The Industrial Revolution in Reverse? *The Planner, 74*, 15–19.

1989 Arcadia for Some: The Strange Story of Autonomous Housing. *Housing Studies, 4*, 149–154.

The Turbulent Eighth Decade: Challenges to American City Planning. *Journal of the American Planning Association, 55*, 275–282.

Strategic Areal Development in Managing Metropolitan Metamorphosis and Development. *Regional Development Dialogue, 10/4*, 1–19.

London 2001. *Revue Française de Civilisation Britannique, 5/3*, 128–137.

The Planners' World. *Futures, 21*, 498–507.

Planning for a Golden Age. *The Planner, 75/29*, 20–24.

Transitional Stress and the Crisis of Those without Hope. *Town and Country Planning, 59*, 331–333.

1991 Three Systems, Three Separate Places. *Journal of the American Planning Association, 57*, 16–19.

Managing Growth in the World's Cities. *Berkeley Planning Journal, 6*, 19–35.

Transit Joint Development in the USA: An Inventory and Policy Assessment. (With J. Landis and R. Cervero). *Environment and Planning C: Government and Policy, 9*, 431–452.

1992 Britain's Cities in Europe. *Town and Country Planning, 62*, 7–13.

Cities in the Informational Economy. *Urban Futures: Issues for Australia's Cities, 5*, 1–12.

Metropoliternes Fremtid. *HUG: Tidsskrift for Kultur og Politik, 62*, 66–72.

Waterfronts, a New Urban Frontier/Waterfront: una Nouva Frontiera Urbana. *Aquapolis, 1*, 6–17.

East Thames Corridor: The Second Golden Age of the Garden Suburb. *Urban Design Quarterly, 43*, 2–9.

Agenda for a New Government. (Proceedings of the Town and Country Planning School). *The Planner, Supplement*, 30–34.

1993 Planning in the 1990s: An International Agenda. *European Planning Studies, 1*, 3–12.

Forces Shaping Urban Europe. *Urban Studies, 30*, 883–898.

Urban Development and Urban Policy: Where have we come from? (Bossom Lectures on Planning for the Future City). *RSA Journal, 141*, 511–522.

The Second Railway Age (with D. Banister). *Built Environment, 19*, 157–162.

1994 Fifty Years of Post-War Planning: Comment on Barry Cullingworth's Paper. *Town Planning Review, 65*, 297–299.

Squaring the Circle: Can we Resolve the Clarkian Paradox? In: *Shape, Rules, and Design: Essays in Honour of the 60th Birthday of the Founding Editor—Lionel March. Environment and Planning B, 21*, s79–s94.

Die Zukunft der Planung. *Mitteilungen der Österreichischen Geographischen Gesellschaft, 136*, 37–52.

On Values and Role Models: The Making of a Planner. *Regenerating Cities, 7*, 12–18.

Time Again for Rail? *Access, 4*, 2–7.

1995 The Future of Cities in Western Europe. *European Review, 3*, 161–169.

Planning Strategies for Cities and Regions. *Town & Country Planning, 64*, 139–142.

Conference Report: 'Seizing the Moment—London Planning 1944–1994'. *The London Journal, 20*, 91–94.

Bringing Abercrombie Back from the Shades: A Look Forward and Back. *Town Planning Review, 66*, 227–241.

Planning and Urban Design in the 1990s. *Urban Design Quarterly, 56*, 14–21.

Planifier les Bassins Métropolitains. *Metropolis, 104/105*, 6–8.

Taming the Monster We Love. *Town and Country Planning, 64*, 295–297.

Planning Strategies for UK Cities and Regions. *Property Review, 5*, 302–305.

1996 Four Million Households—Where Will They Go? (With M. Breheny). *Town and Country Planning, 65*, 39–41.

The Global City. *International Social Science Journal, 147*, 15–23.

London's Wild East: Planning Lessons from Las Vegas. *Planning in London, 16*, 25–28.

It All Came Together in California. *City, 1/2*, 4–12.

Revisiting the Nonplace Urban Realm: Have we Come Full Circle? *International Planning Studies, 1*, 7–16.

So Where *Will* They Go? (With M. Breheny). *Town and Country Planning, 65*, 200–205.

GTVs and Regional Power. *Planning in London, 18*, 10–11.

Where Will the People Go? *Agenda, 1/1*, 13–16.

Cities of People and Cities of Bits. *Demos, 9*, 25–26.

The Geography of the Information Economy. *Interdisciplinary Science Reviews, 21*, 199–208.

Where Will They Go? A Response to the Response. *Town and Country Planning, 65*, 290–291. (With M. Breheny).

1946–1996—From New Town to Sustainable Social City. *Town and Country Planning, 65*, 295–297.

Ealing, The Queen of London's Victorian Railway Suburbs. *Urban Design Studies, 2*, 31–44.

1997 Establishment Man or Maverick? Interview with Peter Hall. *City, 5–6*, 159–173.

Die europäische Stadt: Vergangenheit und Zukunft. *Die Alte Stadt, 1/97*, 18–34.

Face the Future—Just like Ebenezer Howard. *Planning in London, 21*, 14–17.

The Future of the Metropolis and its Form. *Regional Studies, 31*, 211–220.

Reflections Past and Future on Planning Cities. *Australian Planner, 34*, 83–89.

Le New Towns in Gran Bretagna: Passato, Presente e Futuro. *Urbanistica, 107*, 141–145.

The University and the City. *GeoJournal, 41*, 301–309.

Modelling the Post-Industrial City. *Futures, 29*, 311–322.

The Urban Challenge. *UK CEED Bulletin, 40*, 14–15.

The Next 50 Years. *Town and Country Planning, 66*, 285–289.

The Great Housing Land Debate. *Parliamentary Review—Town Planning Review*, Winter 1997, 42–43.

1998 *To-Morrow* for Today. *Town and Country Planning, Special Supplement*, 2–3.

The Three Magnets re-interpreted. *Town and Country Planning, Special Supplement*, 68.

"A Telegram from the Queen": The Centenary of Modern Planning. *Planning History, 20/3*, 8–11.

1999 Transport Corridors for the Randstad. *Stedebouw & Ruimtelijke Ordening, 79/3*, 23–27.

I Motori della Concorrenza Economica Globale. *Urbanistica, 111*, 55–57.

How Cities Can be Expected to Change. *Australian Planner, 36*, 66–71.

The Future of Cities. *Computers, Environment and Urban Systems, 23*, 173–185.

Sustainable Cities or Town Cramming. *RSA Journal, 148/4*, 72–80.

Creative Cities. *Cities and Design: An Academic Journal for Intercity Networking, 7/8*, 1–24. (In Mandarin).

2000 The Dumills—Londoners 2000. The View from 1963, and the 1963 View from 2000. *Town and Country Planning, 69*, 44–46.

Creative Cities and Economic Development. *Urban Studies, 37*, 639–649.

Urban 21: Global Conference in Berlin. *Deutschland, 4/2000*, 24–30.

Planning and Project in the Present-Day City. *Cartas Urbanas, 6*, 86–97.

Cidades Inovadoras. *Desenvolvimento Urbano e Meio Ambiente, 40*, 1–4.

Planifación y Gestión de la Ciudad para la Sociedad Emergente. *Urban, 4*, 14–30.

The Grid as Generator. *Arq—Architectural Research Quarterly, 4*, 309–10.

2001 L'Invité: Peter Hall. *Urbanisme, 315*, 14–22.
High-Speed Rail Comes to London. *Access, 19*, 26–31.
The Changing Urban Hierarchy in England and Wales, 1913–1998 (with S. Marshall and M. Lowe). *Regional Studies*, 35, 775–807.
Who's Up and Who's Down in the Urban Hierarchy. *Town and Country Planning, 70*, 268–9.
2002 Knowledge Society and 21st Century Planning. (Interview with So-Sik, Yang). *Planning and Policy (Korea), 2005/09,* 81–91.
Blowing Up Bridges. *Town and Country Planning, 71*, 51.
New Towns—Just History, or an Idea for Today? (with N. Schoon). *Town and Country Planning, 71*, 189–191.
Growth Area Studies and North Sea Fog. *Town and Country Planning, 71*, 258–259.
2003 The End of the City? *Alpbacher Architektgespräche 2002 (Albach Architecture Forum 2002 Discussion),* n.p. Albach: Europäisches Forum Albach.
The End of the City? "The Report of My Death was an Exaggeration". *City, 7*, 141–152.
Jean Gottmann 1915–1994. (Biographical Memoirs of Fellows, II). (With Clout, H.) *Proceedings of the British Academy, 120*, 201–215.
Cities in Civilization: Culture, Innovation and Urban Order. *Journal of Irish Urban Studies, 2/2*, 1–14.
2004 The Buchanan Report: 40 years On. *ICE: Proceedings of the Institution of Civil Engineers, 157*, 7–14.
Walking the Walk. (RTPI Annual Lecture 2003). *Town and Country Planning, 73*, 54–59.
THEME: The Papers of Peter Hall. (In Mandarin). *Urban Planning Overseas*, 19/4, 4–34.
The Global City. *Urban Planning Overseas, 2004/4*, 6–16 (in Mandarin).
Creativity, Culture, Knowledge and the City. *Built Environment, 30*, 256–258.
Multifunctional Urban Planning of Mega-City-Regions. (With Priemus, H.) *Built Environment, 30,* 338–349.
2005 Making Places of Delight. *Town and Country Planning, 74*, 14–15.
Towards Joined-up Planning—Making a Start in West London (with N. Falk). *Town and Country Planning, 74*, 128–131.
Are America and Europe Alien Worlds in Planning? *Harvard Design Magazine, 22*, 87–94.
Knowledge Society and 21st Century Urban Planning. *Planning and Policy (Korea Institute for Human Settlements), 2005/09*, 83–91.
La Nuova Era del Treno. Implicazioni per città e Regioni. *Parametro, 258–259*, 32–35.
Apologia Pro Vita Sua. *Balzan Prize Committee Lecture.*

2006 Changing Travel to Work Patterns in South East England (With Titheridge, H.). *Journal of Transport Geography, 14,* 60–75.
Re-Urbanizing the Suburbs? The Role of Theatre, The Arts and Urban Studies. (With Hall, P.) *City, 10,* 377–392.
A New Geography of Britain: Analects for a Non-Autobiography. *Town and Country Planning, 75*, 304–308.

2007 Des *Docklands* à *Thames Gateway*: Rééquilibrer le Développement de Londres vers l'Est. *Cahiers de l'Institut d'Aménagement et d'Urbanisme de la Région d'Ile-de-France, 146,* 39–46.
New East Manchester: Urban Renaissance or Urban Opportunism? (With Mace, A., Gallent, N.) *European Planning Studies* 15(1), 51–65.
Beyond the Automobile? *Access, 30,* 10–15.
Melvin M. Webber: Maker and Breaker of Planning Paradigms. *Access, Special Issue, Winter 2006–7*, 17–22.

2008 New New Towns. *Blueprint, 263*, 46–49.
From Metropolis to Polyopolis (with Pain, K.) *Urban Planning International, 23,* 15–27.
From Strategy to Delivery: Policy Responses (with Pain, K.) *Urban Planning International, 23,* 28–40.
Comment on Historical ICE Paper. *Urban Planning and Design* (*Proceedings of the Institution of Civil Engineers, 161*), *DP1*, 23–26.
Informational Quantity versus International Quality: The Perils of Navigating the Space of Flows (with Pain, K.). *Regional Studies, 42,* 1065–1077.
London Voices, 1957–2007: From Family and Kinship to London Lives. *City* 12, 50–63
Key Issues for Planning Futures and the Way Forward (with Davoudi S. and Power A.) *Twenty First Century Society*, 3, 229–247

2009 Magic Carpets and Seamless Webs: Opportunities and Constraints for High Speed Trains in Europe. *Built Environment, 35,* 59–69. ISSN 0263-7960.
Chinese Urban Growth in a Global Perspective. *Urban Planning International, 24,* 9–15. ISSN 1673-9493.
Gathering Evidence for Collective Reformation. *Town and Country Planning, 78,* 152–154. ISSN 0040-9960.
Looking Backward, Looking Forward: The City Region of the Mid-21st Century. *Regional Studies, 43,* 803–817. ISSN 0034-3404.
The Polycentric Metropolis: A Western European Perspective on Mega-City Regions. *Journal of Urban and Regional Planning, 2/3, Series 6*, 1–17. ISBN 978-7-100-06733-1.

2011 Die Zukunft der Städte im Zeitalter der wissensbasisierten Ökonomie. In: *RegioPol: Zeitschrift für Regionalwirtschaft*, 1+2 2011, *Urbane Zukunft in der Wissensökonomie*, 35–46. ISSN 2192-1601.

The impacts of High-Speed Trains on British Economic Geography: a Study of the UK's InterCity 125/225 and its Effects. (with Chen C-L.) *Journal of Transport Geography, 19,* 689–704. ISSN 0966-6923.

High-Speed Gateways. (with Wray I.) *Town and Country Planning, 80,* 322–327. ISSN 0040-9960.

2012 New Perspectives on Transportation and Urban Density. *Urban Solutions, 1,* 32–37. ISSN 2301-3540.

All change in the Lower Lea Valley. *Australian Planner, 49,* 193–194. ISSN 0729-3682.

The wider spatial-economic impacts of high-speed trains: a comparative case study of Manchester and Lille sub-regions. (with Chen C-L.). *Journal of Transport Geography, 24*, 89–110. ISSN 0966-6923.

Drawing on the Garden City Prescription. *Town and Country Planning*, 82 (11), 475–478

2013 Using HS2 to irrigate the Regions (with Chen C-L.) *Town and Country Planning*. 82(4) 170–174

Refreshing the Parts that Other Transport cannot Reach. *Town and Country Planning*. 82(3), 121–132

2014 We have a problem—so what to do? *Town and Country Planning*, 114–117

High-speed north: Building a Trans-pennine Mega-city (with Thrower D. and Wray I.) *Town and Country Planning* 83(4), 160–167

Apologia pro vita sua. *Town and Country Planning*, 317–325

The International Geographical Union

The International Geographical Union (IGU) is an international, non-governmental, professional organization devoted to the development of the discipline of Geography. The purposes of the IGU are primarily to promote Geography through initiating and coordinating geographical research and teaching in all countries of the world. Its work is conducted through the instruments of its National Committees, Commissions and Task Forces. The IGU hosts the International Geographical Congress every four years and also promotes regional conferences and other meetings that further of the objectives of the Union. The IGU also facilitates the participation of geographers in the global community of scientists through its formal affiliation as a Member Union within both the International Council for Science (ICSU) and the International Social Science Council (ISSC).

The IGU was formally established in Brussels in 1922. However, the history of international meetings of geographers is much longer. The first of a series of congresses met in 1871 in Antwerp. Since its early days the Union has consisted of three major components: a General Assembly of the delegates appointed by the member countries which meets at the time of the Congress and is the highest authority of the Union; an Executive Committee which consists of a President, eight Vice-Presidents and a Secretary-General and Treasurer; Commissions and Study Groups which continue their work between General Assembly meetings. The working languages of the Union are English and French.

Website: http://igu-online.org/.

R.D. Knowles and C. Rozenblat (eds.), *Sir Peter Hall: Pioneer in Regional Planning, Transport and Urban Geography*, SpringerBriefs on Pioneers in Science and Practice 52, DOI 10.1007/978-3-319-28056-1

About the Co-editors

Professor Richard D. Knowles (UK) BA, Ph.D. is Emeritus Professor of Transport Geography, University of Salford. He is President and Chair of the International Geographical Union's Commission on Transport and Geography. He was the Founding Editor of the *Journal of Transport Geography* (published by Elsevier) between 1993 and 2012. Professor Richard Knowles' research and publication focus on two main themes:

- Examining effects of transport policy changes, including bus deregulation and privatization, and rail franchising
- Analysing transport and spatial development impacts of new transport infrastructure and technology at global and local levels, especially fixed links, urban rail transit and transit-oriented development in the UK and Scandinavia.

He is a specialist in Transport Geography and in Scandinavian Human Geography and has published several key books and reports and numerous research journal papers and book chapters. He was also Co-Editor of Ashgate's *Transport and Mobility* Book Series. Professor Knowles received the Edward Ullman Award 2004 (Association of American Geographers), and the Alan Hay Award 2010 (Royal Geographical Society), for significant contributions to Transport Geography.

Among his selected publications are: Fullerton B. and Knowles R.D. (1991) *Scandinavia*, Paul Chapman; Hoyle B.S. and Knowles R.D. (eds.) (1992, 1998) *Modern Transport Geography*. Belhaven; Wiley; Knowles R.D. (2006) Transport shaping space: differential collapse in time-space. *Journal of Transport Geography*, 13, 407–425; Knowles R.D. (2013) Railway franchising in Great Britain and the effects of the 2008–2009 economic recession, *Environment and Planning A*, 45(1), 197–216; Knowles R.D., Shaw J. and Docherty I. (eds.) (2008) *Transport Geographies: Mobilities, Flows and Spaces*. Blackwell.

R.D. Knowles and C. Rozenblat (eds.), *Sir Peter Hall: Pioneer in Regional Planning, Transport and Urban Geography*, SpringerBriefs on Pioneers in Science and Practice 52, DOI 10.1007/978-3-319-28056-1

Address: School of Environment & Life Sciences, Peel Building, University of Salford, Salford M5 4WT, Greater Manchester, United Kingdom.
Email: r.d.knowles@salford.ac.uk.

Professor Céline Rozenblat (France/Switzerland), Ph.D. Her research focuses on the effects of multi-national firms networks on cities systems' evolution in multi-level territorial processes. She has organized workshops for high-level policy makers (French IHEDAT), and produced numerous comparative studies of cities for planning institutions such as DATAR in France or ESPON in Europe. She participates currently in the FET MULTIPLEX FP7, a project led by physicists on complex networks and she leads the FNS LOGIICS project on the integration of Chinese and Indian cities in the globalization process. Professor Rozenblat is President of the Urban Commission of the *International Geographical Union* (IGU) and coordinates the Territorial Intelligence Dep. Of the Utwin UNESCO Digital Campus of Complex Science. She serves on the Executive Committee of the Complex Science Society.

Among her selected publications are: Rozenblat C., Melancon (eds.) (2013) *Multilevel analysis and visualisation of geographical networks.* Springer, Methodos Series, No. 11; Hu Y., Zhou D., Zhang R., Han Z., Rozenblat C. and Havlin S. (2013) Percolation of interdependent networks with intersimilarity. *Physical Review E*, 88(5), 052805; Garas A., Argylakis P., Rozenblat C., Tomassini M. and Havlin S. (2010) Worldwide spreading of economic crisis. *New Journal of Physics*, 12, 113043, arXiv:1008.3893v1 [physics.soc-ph].

Address: Institute of Geography and Sustainability, University of Lausanne, Geopolis Building Mouline area, CH-1015 Lausanne, Switzerland.
Email: Celine.rozenblat@unil.ch.

About the Contributors

Professor Michael Batty (UK) is Bartlett Professor of Planning and Chair of Centre for Advanced Spatial Analysis (CASA). He was Peter Hall's Research Assistant, Urban Systems Research Unit, Department of Geography, University of Reading 1969–1972 and Peter Hall's Departmental Colleague, University of Reading 1972–79.

Address: University College London, 99 Tottenham Court Road, London W1N 6TR, UK.
Email: m.batty@ucl.ac.uk.

Dr. Chia-Lin Chen (China) is Research Associate in Transport and Development, University College London, until summer 2015; Lecturer in Urban Planning and Design, Xi'an Jiaotong Liverpool University, Suzhou, China from autumn 2015. She was the last (of many) Ph.D. students supervised by Peter Hall and was research collaborator on Peter Hall's EU SINTROPHER project.

Address: University College London, Central House, 14 Upper Woburn Place, London WC1H 0NN, UK.
From Autumn 2015: Urban Planning and Design, Xi'an Jiaotong Liverpool University, Suzhou, China.
Email: chia-lin.chen@alumni.ucl.ac.uk; chia-lin.chen@xjtlu.edu.cn.

Dr. Dan O'Donoghue (UK) is Principal Lecturer in Geography and Vice President of the International Geographical Union's Urban Commission.

Address: University of Canterbury in Kent, UK.
Email: daniel.donoghue@canterbury.ac.uk.

Professor Kathy Pain (UK) is Professor in Real Estate Development and Director of Research. She was one of the early Ph.D. students supervised by Peter Hall, researching into housing markets and urban policy. Co-Director with Peter Hall on the POLYNET research project.

Address: School of Real Estate & Planning, University of Reading, Whiteknights, Reading RG6 6UD, UK.
Email: k.pain@reading.ac.uk.

R.D. Knowles and C. Rozenblat (eds.), *Sir Peter Hall: Pioneer in Regional Planning, Transport and Urban Geography*, SpringerBriefs on Pioneers in Science and Practice 52, DOI 10.1007/978-3-319-28056-1

Dr. Jonathan Reades (UK) is Lecturer in Quantitative Human Geography. One of the later PhD students supervised by Peter Hall, researching into communications data for the POLYNET research project.

Address: Department of Geography, Kings College London, Strand Campus, strand, London WC2R 2LS, UK.
Email: jonathan.reades@kcl.ac.uk.

Professor Peter J. Taylor (UK) is Professor of Human Geography, Northumbria University, Newcastle upon Tyne; Emeritus Professor, Department of Geography, Loughborough University; Director of GaWC (Globalization and World Cities Research Network).

Research collaborator with Peter Hall on the POLYNET research project.

Address: Faculty of Engineering and Environment, Northumbria University, Newcastle upon Tyne NE1 8ST, UK.
Email: crogfam@yahoo.com.

About this Book

Sir Peter Hall's immense contribution to the understanding of city functions, planning, structures, regeneration and futures was built upon inter-disciplinary research of the highest international quality, rooted in his training, reading and experience in geography and planning. This book offers a unique insight into Professor Peter Hall's pioneering ideas, vision and inter-disciplinary methods to practitioners, policy makers and academic researchers in regional planning, transport and urban geography, and urban regeneration, throughout the world.

This book specifically offers:

- Deep insights into a range of Peter Hall's historical, contemporary and on-going global and UK research projects
- an excellent summary of Peter Hall main visions of urban futures and his contributions
- current major researchers of international renown in urban and transport geography, planning and real estate explaining their research and their link to Peter Hall's visions
- joint research between Peter Hall and five of his research collaborators, three of whom he supervised for doctoral research
- a complete bibliography of Peter Hall's immense output of published research sustained for 55 years from 1960 to 2015 in hundreds of books, journal papers and articles.

The *International Geographical Union's* (IGU) Commission on Transport and Geography together with the IGU Urban Commission have produced this book to both commemorate and further publicise Sir Peter Hall's immense contribution to knowledge. Specialists of both geography and planning disciplines testify to Peter Hall's multiple contributions, essentially in regional and urban planning, urban geography and transport geography.

After an Introduction to Peter Hall's immensely successful international career by Richard Knowles and Celine Rozenblat, Peter Taylor presents Peter Hall's high-level contribution as "a polymath in city studies". Jonathan Reades investigates Peter Hall's research into "Location and Innovation". Peter Hall's research into "Transport and Place-Making: a Long View" is examined by Chia-Lin Chen.

R.D. Knowles and C. Rozenblat (eds.), *Sir Peter Hall: Pioneer in Regional Planning, Transport and Urban Geography*, SpringerBriefs on Pioneers in Science and Practice 52, DOI 10.1007/978-3-319-28056-1

Kathy Pain explains her global research with Peter Hall entitled "The Strategic Planning Protagonist: Unveiling the Global Mega-City Region", whilst Michael Batty recalls how he collaborated with Peter Hall on research into "Creative Destruction, Long Waves and the Age of the Smart City". Celine Rozenblat and Dan O'Donoghue conclude by examining Peter Hall's standing as "The Visionary of World and European cities".

More on this book is at: http://afes-press-books.de/html/SpringerBriefs_PSP_SirPeterHall.htm.

MIX
Papier aus verantwortungsvollen Quellen
Paper from responsible sources
FSC® C105338

If you have any concerns about our products,
you can contact us on
ProductSafety@springernature.com

In case Publisher is established outside the EU,
the EU authorized representative is:
Springer Nature Customer Service Center GmbH
Europaplatz 3, 69115 Heidelberg, Germany

Printed by Libri Plureos GmbH
in Hamburg, Germany